golf science

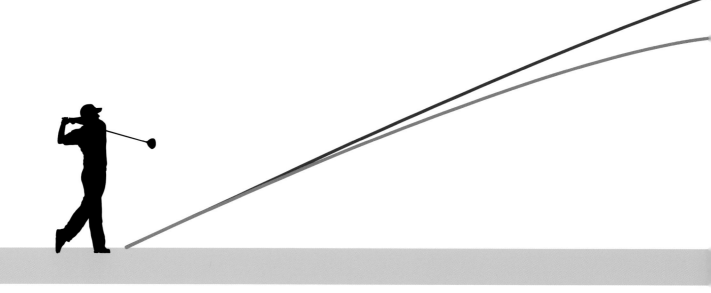

golf science

Edited by **MARK F. SMITH**

optimum performance from tee to green

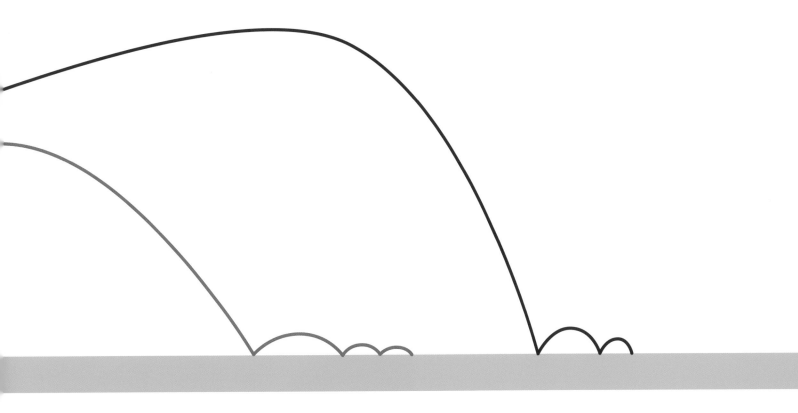

THE UNIVERSITY OF CHICAGO PRESS

Chicago and London

The University of Chicago Press, Chicago 60637
The University of Chicago Press, Ltd., London
© The Ivy Press Limited 2013
All rights reserved. Published 2013.
Printed in Singapore

A CIP record for this title is available at the Library of Congress.

22 21 20 19 18 17 16 15 14 13   1 2 3 4 5
ISBN-13: 978-0-226-00113-5 (cloth)
ISBN-10: 0-226-00113-X (cloth)
DOI: 10.7208/chicago/9780226001272.001.0001

⊛ This paper meets the requirements of ANSI/NISO
Z39.48-1992 (Permanence of Paper).

Color origination by Ivy Press Reprographics.

This book was conceived, designed, and produced by

Ivy Press
210 High Street, Lewes
East Sussex BN7 2NS
United Kingdom
www.ivypress.co.uk

Creative Director Peter Bridgewater
Publisher Jason Hook
Editorial Director Caroline Earle
Art Director Michael Whitehead
Designer Lisa McCormick
Commissioning Editor Kate Shanahan
Project Editor Rob Yarham
"Science in Action" text Mark F. Smith and Rob Yarham
"Equipment" spread text Mark F. Smith
Illustrator Nick Rowland
Additional illustrations Alan Osbahr

Cover illustration: Nick Rowland; photo: Getty Images/Phil Leo/Michael
Denora

Dedication
This book is dedicated
to my beautiful children
D & C.
"The best preparation for
tomorrow is just doing your
very best today."

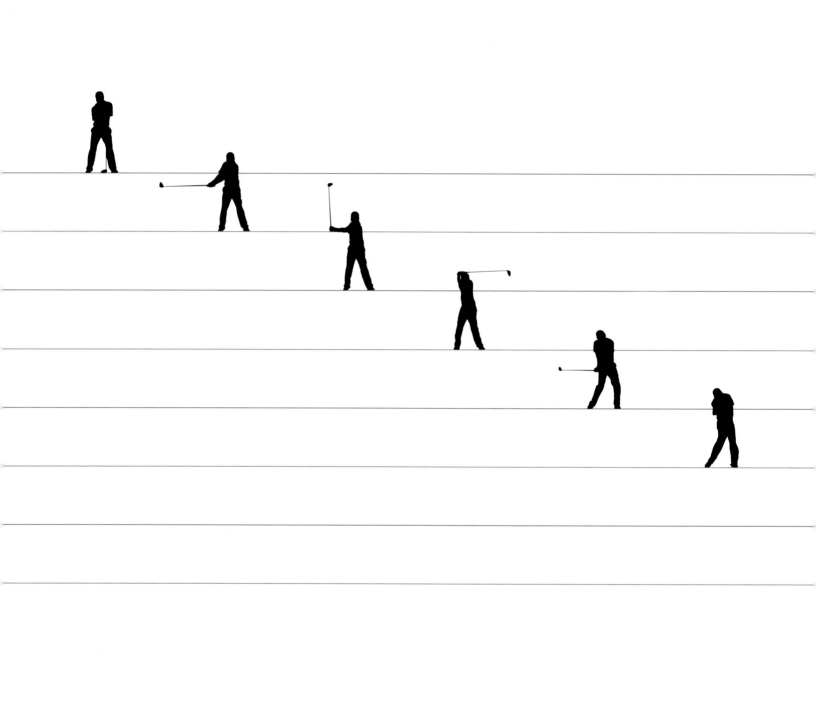

Contents

Introduction

Playing golf can be one of the most simple yet perplexing of pastimes. For the millions of players around the world, who challenge themselves each time they pick up a club, golf offers the chance to escape among beautiful scenery, with shimmering green fairways and carefully planted shrubbery and flowers. Golf challenges our mind, as well as body, in ways that can bring elation and despair in the space of a moment. The act of getting a ball in a hole is simply complex when one ponders the number of factors that all interact with the small white spherical object: the golfer, the equipment, the course, and the weather. Understanding the science behind golf enhances both the interest in, and the sheer enjoyment of, playing the game.

From across the globe *Golf Science* brings to you leading experts in the physical, mental, technical, and tactical aspects of the game. This book binds together their significant findings with those of scientists who have parsed and studied every element of the game. Some discoveries were made in the nineteenth century and others as recently as this year. A wide range of sciences is embraced in the book because golf involves a surprisingly diverse array of disciplines.

The fundamental principles of physics lead to the magic of engineering and technology, which has led to the development of innovative club design, futuristic coaching devices, and novel practice approaches. There are the mysteries of biomechanical forces, as well as crucial explanations of aerodynamics. You, the golfer, are also dissected and put under the microscope to reveal how the connection between your mind and body allows you to repetitively execute the complex task of hitting a ball.

This is not a manual nor an instruction guide. It goes deeper than such books, but it will help everyone to get more from their golf game—whether picking up a club for the first time, hacking balls down the range,

▶ **Science behind success** *Even a leading Tour pro like Rory McIlroy runs into trouble sometimes. Professional golfers have turned increasingly to science to help them understand every aspect of the game and give them that edge out on the course. Psychology, physiology, nutrition, materials science, and physics all play their part in the golfer's quest to maximize their chances of success, even when things don't quite go according to plan.*

navigating their way around 18 holes, or competing in tournament play to win. Golf is about the breeze on your brow, watching the ball fly toward the hole, knowing that your swing felt right. Without a doubt, your potential to do so can be boosted by your knowledge of the science at play.

Scientific discoveries are not always easy to comprehend, but this book presents them in a straightforward way. The *Question-and-answer* pages contain unique info-graphics that convey scientific results very clearly. Every science question is mirrored by a question posed by a golfer. The text and the info-graphics combine to answer them both.

The *Equipment* pages show how and why crucial elements of the game have helped revolutionize the way we play golf. From early sixteenth century club-like sticks carved from a single piece of wood to highly sophisticated, scientifically-inspired precision tools, golf has embraced the scientific revolution.

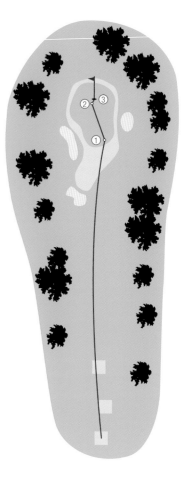

▶ **Graphically speaking** *The graphics and illustrations in the book will introduce you to the science that has developed over several hundred years of golfing, from the fundamental physics and body biomechanics involved when executing a golf swing to the latest technical developments in golfing equipment.*

Spin loft

Club direction

It is all very well explaining the what, why, how, and when, but, apart from playing the courses, little can bring a golfer closer to their favorite activity than a world-class photograph. So, the photographic *Science in action* feature pages reveal how the principle is applicable to the practice, while at the same time conveying the drama, excitement, and human endeavor that is golf.

For golfers who are eager to delve deeper, there is a comprehensive glossary that defines many of the terms and concepts. An extensive list of references points readers to the sources of the information, so that it is easy to engage more closely with the science.

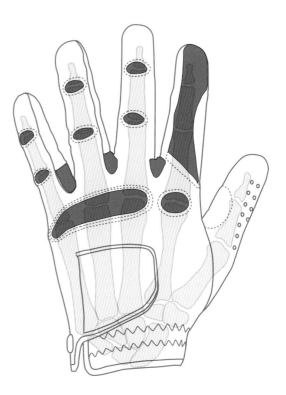

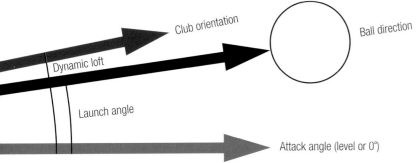

Club orientation

Ball direction

Dynamic loft

Launch angle

Attack angle (level or 0°)

Without question the art and science of golf primarily takes place in the mind, played out through the way we move our bodies. This instinctive connection affects how we feel, how we think and then how we swing our golf club. There is no better place to illustrate this than on the course, where our golf can be dramatically affected by what happens in our mind. Tour players have an amazing ability to regulate their thoughts while still allowing their bodies to function at a very high level. How body biomechanics, anatomy, physiology, and mental strategies all link together is revealed throughout this chapter. Mark F. Smith explores, through a series of intriguing questions, the science behind the mind–body connection and how it relates to the way the golfer moves and feels, interacting with the equipment and creating movement of the body and ball.

chapter one

mind and body

Mark F. Smith

Does physical fitness level affect performance?

Will getting fit improve my game?

From a less-informed perspective, success in golf is often seen to be more about technical, tactical, and mental factors than physical ones. Indeed it is true that, historically, physical fitness has not appeared to be of that much importance in golf. Today, however, even some of the "old school" pros are catching on. Miguel Angel Jiménez goes for a jog every morning, and Open champion Darren Clarke has revealed that more time spent working on his fitness has been a key factor in his success. Research proves that good physical attributes—especially stamina, strength, mobility, and balance—help to improve golfing performance and lower those scores.

Over the short term, walking and golf-related training has been shown to elicit a number of both general health and golf-specific performance improvements. A reduction in body fat helps to maintain a healthy blood pressure and increases functional capacity, both factors associated with the reduced risk of hypokinetic diseases. In other words, undertaking regular golfing activities in conjunction with a healthy lifestyle will decrease your chances of developing serious illnesses and will also increase your physical abilities and life expectancy.

Evidence also reveals that performing a few simple exercises each day will increase your strength, mobility, and balance. Assessing more than 250 players, with a handicap range between 0 and 20, a 2009 research review revealed that playing standard is related to trunk, hip, and shoulder strength—the lower the handicap, the stronger the player was in these critical areas.[1] Additionally, a number of other studies have identified a link between mobility in the hip, mid-trunk, and shoulder with increased clubhead speed—which means more carry on long shots and greater spin on short shots.[2] Studies show that adopting carefully managed physical conditioning routines—such as flexibility, balance, and strength exercises—at least three times per week for 8–10 weeks has a positive effect on clubhead speed, whatever a golfer's ability.

▶ *Healthy body, healthy golf* *According to the latest golf science research, technical, tactical, mental, and life skills will affect—but will also be influenced by—the physiological status of the player.[3,4] Prior to play, the golfer's objective should be to ensure that their aerobic fitness, strength and conditioning, flexibility, podiatric and optometric performance, dietary habits, and injury management are all factored into their overall player plan. Once play begins, the player must then select appropriate strategies, bearing in mind external factors, such as temperature and humidity, in order to maximize their performance—proper nutrition, hydration, clothing, physical preparedness, and mental focus.*

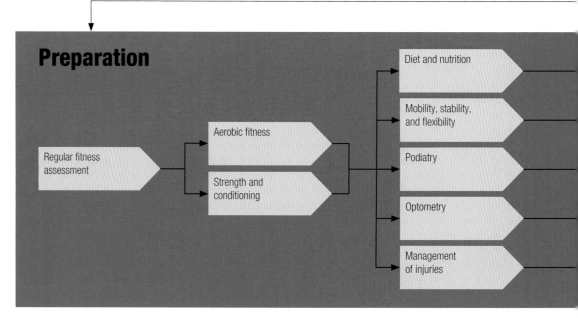

Preparation

Regular fitness assessment → Aerobic fitness / Strength and conditioning → Diet and nutrition / Mobility, stability, and flexibility / Podiatry / Optometry / Management of injuries

Shaping up

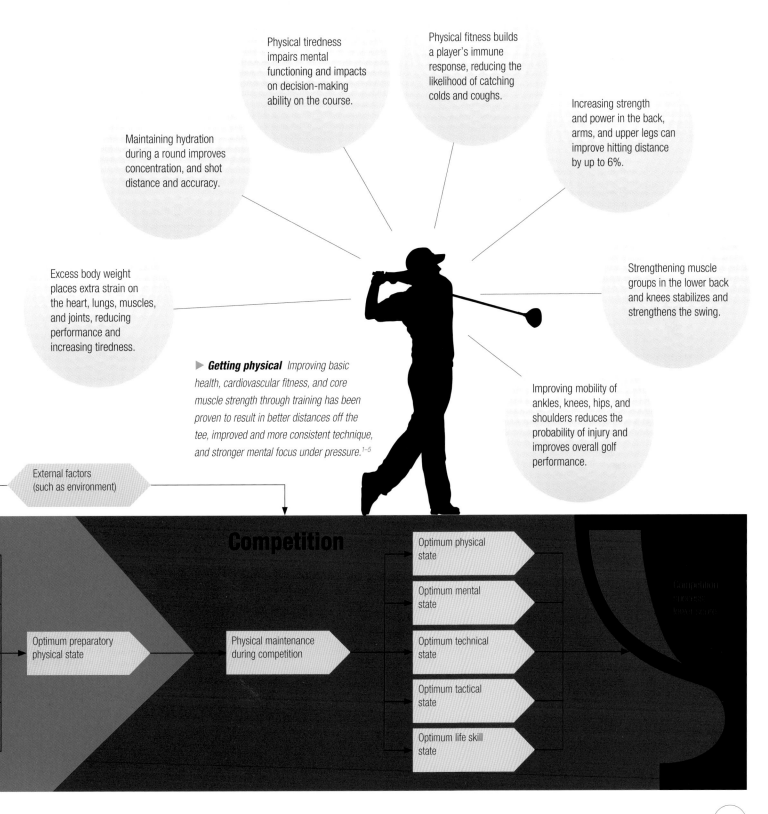

Physical tiredness impairs mental functioning and impacts on decision-making ability on the course.

Physical fitness builds a player's immune response, reducing the likelihood of catching colds and coughs.

Maintaining hydration during a round improves concentration, and shot distance and accuracy.

Increasing strength and power in the back, arms, and upper legs can improve hitting distance by up to 6%.

Excess body weight places extra strain on the heart, lungs, muscles, and joints, reducing performance and increasing tiredness.

Strengthening muscle groups in the lower back and knees stabilizes and strengthens the swing.

▶ **Getting physical** *Improving basic health, cardiovascular fitness, and core muscle strength through training has been proven to result in better distances off the tee, improved and more consistent technique, and stronger mental focus under pressure.*[1–5]

Improving mobility of ankles, knees, hips, and shoulders reduces the probability of injury and improves overall golf performance.

External factors (such as environment)

Competition

Optimum preparatory physical state

Physical maintenance during competition

Optimum physical state

Optimum mental state

Optimum technical state

Optimum tactical state

Optimum life skill state

Do "quiet eye" moments help putting success?

Where should I look when making a putt?

Where a golfer looks during the putt may reveal more about what's happening in the mind during those vital seconds beforehand than first thought. Scientists from Canada and the United Kingdom have uncovered the role our brain's visual motor control system may play in enhancing neurological efficiency throughout the stroke.[1-4] More importantly, though, they have revealed a solution which helps all players, irrespective of ability, improve their putting performance.

Evidence presented at the 2012 World Scientific Congress of Golf Science reveals that focusing on the ball in a particular way—dubbed "quiet eye" moments—eliminates unwanted distractions, and leads to more successful putting. Based on a number of controlled experimental studies, it has been suggested that the key is to spend around two seconds during the stroke concentrating on the ball and then, once impact has occurred, to continue staring at the same spot on the ground afterward.

It is thought that this approach is effective because it allows the golfer to take in only the necessary visual information required to make the shot. Focusing away from the intended task at hand can disrupt the functioning of millions of neurons in the brain that convert the visual information into movements of the putter. Given that putting is a hugely important part of golf, accounting for around 45 percent of the shots taken in an average round, researchers are beginning to acknowledge that this approach may be vital to success, improving both precision and accuracy, and preventing the breakdown of the movement under high levels of pressure and nerves.

▼ ***"Quiet eye" study*** *Golfers were monitored on a follow-up trial and another performed under pressure. Those who undertook "quiet eye" training showed marked increases in the time spent fixating on the target line and location of the ball, and, more importantly, the percentage of putts holed from 10 ft (3 m) improved.*

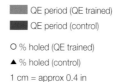

- ▮ QE period (QE trained)
- ▮ QE period (control)
- ○ % holed (QE trained)
- ▲ % holed (control)
- 1 cm = approx 0.4 in

▼ ***Putt it there*** *By adopting the stance, then gazing calmly and steadily at the hole, bringing the eyes back to the ball and then fixating on the back on the ball, the golfer should be able to get their longer putts much closer, avoiding those embarrassing three-putt moments.*

- ▮ Performance error (QE trained)
- ▮ Performance error (control)
- 1 cm = approx 0.4 in

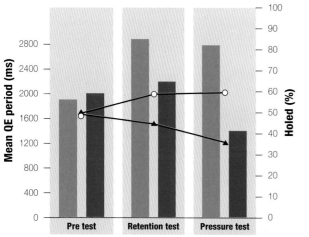

Study results

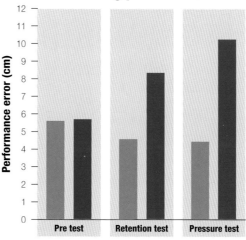

Putting performance

Eyes on the prize

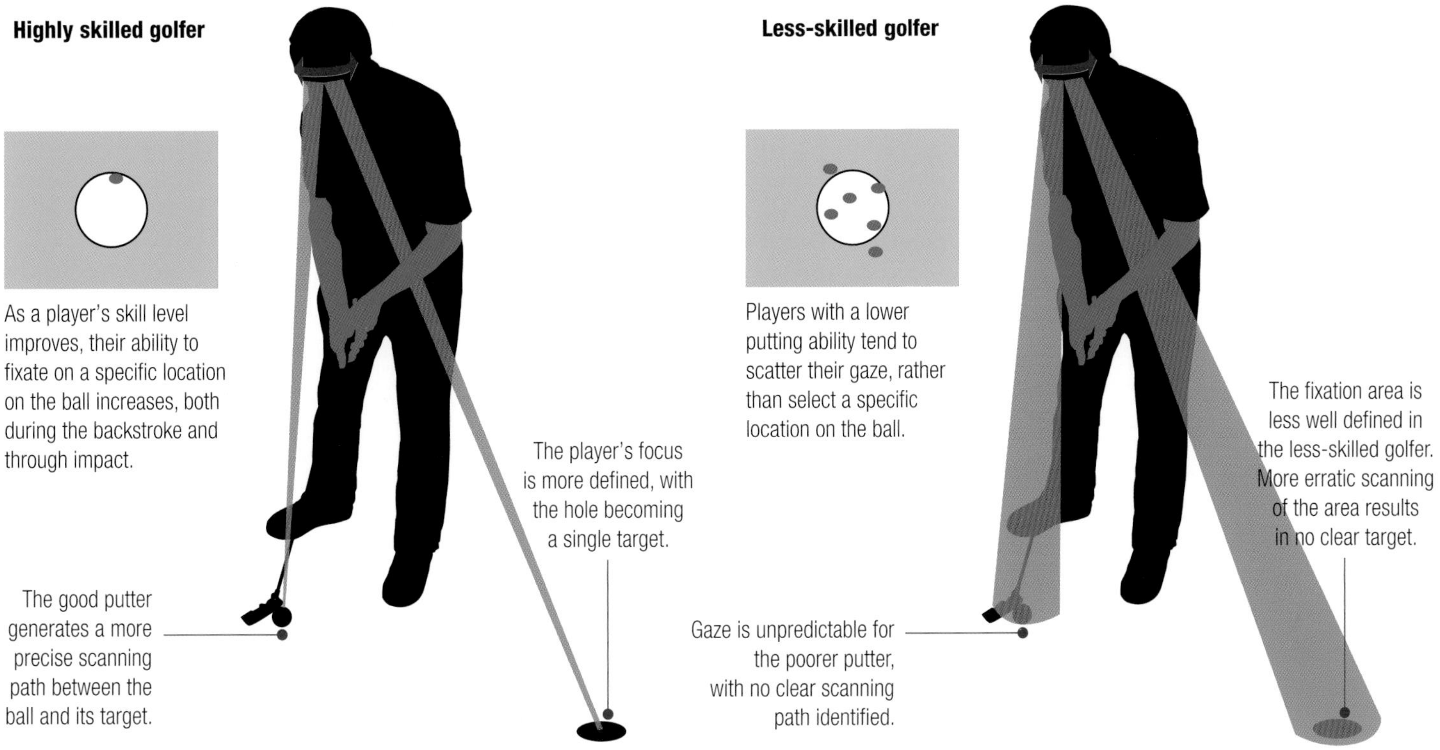
Highly skilled golfer
As a player's skill level improves, their ability to fixate on a specific location on the ball increases, both during the backstroke and through impact.
The good putter generates a more precise scanning path between the ball and its target.
The player's focus is more defined, with the hole becoming a single target.
Less-skilled golfer
Players with a lower putting ability tend to scatter their gaze, rather than select a specific location on the ball.
Gaze is unpredictable for the poorer putter, with no clear scanning path identified.
The fixation area is less well defined in the less-skilled golfer. More erratic scanning of the area results in no clear target.

▲ ***On the ball*** *Analyzing a golfer during the final moments of a putt has revealed what may be two key indicators of putting success: where the player fixes their gaze prior to making the backswing; and how long they stare at a particular location on the ball. In a series of studies investigators found that highly-skilled golfers focused on a precise point on the ball and target line.*[1–3] *It is thought that the brain's neural networks are able to initiate a good putt during these "quiet eye" moments, while overriding competing neural processes responsible for generating distractions and anxiety.*

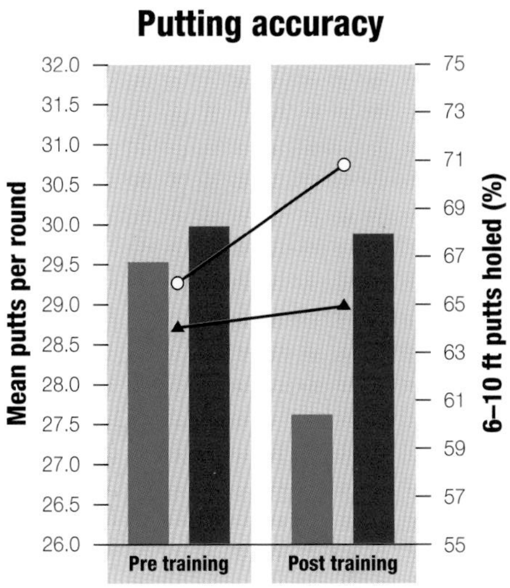
Putting accuracy
Mean putts per round
32.0
31.5
31.0
30.5
30.0
29.5
29.0
28.5
28.0
27.5
27.0
26.5
26.0
6–10 ft putts holed (%)
75
73
71
69
67
65
63
61
59
57
55
Pre training
Post training
○ % 6–10 ft holed (QE trained)
▲ % 6–10 ft holed (control)
Putts per round (QE trained)
Putts per round (control)

◀ ***Quiet time*** *"Quiet eye" training seems to improve putting scores during competitive on-course play.*[2] *Performance data collected over a number of competitive rounds before and after training revealed that players who embarked on a regular training regime reduced their number of putts per round by an average of 1.92 shots, holing out 5 percent more putts from 6–10 ft (2–3 m). Interestingly, when compared to US PGA 2011 Tour putting statistics, if a golfer ranked last (186th) experienced an improvement of 1.92 putts per round, they would climb 139 places to 47th in the rankings.*

Need to know

The "quiet eye" technique:

- Line up a shot, alternating with quick fixations between ball and hole.
- Before and during the stroke, hold a steady fixation on the back of the ball, for around 2–3 seconds.
- After contact with the ball, keep the eyes steady for a further half a second.

Does aging impact on golf performance?

As a senior, does golf improve or damage my health?

An inevitable part of life is the fact that we all grow old. How our bodies age is a hotly disputed topic among gerontologists (the scientists who study the aging process). Some suggest that aging is largely the consequence of a series of random events, experienced through our interaction with the environment, altering and eventually damaging our molecular make-up. Others conclude that random events alone are not sufficient to explain all aging processes. They believe that aging is more a matter of destiny and that our lifespans are in part pre-programmed even before our births. Whatever the complex mix of genetic, environmental, lifestyle, and socioeconomic factors influencing the lifelong process, we can be certain that aging results in a progressive loss of physical and mental function.[1]

It's not all bad news for the aging golfer. Despite many of the age-related changes affecting older players' risk profiles, playing golf regularly offers ways of preserving flexibility, strength, endurance, muscular speed, balance, and cognitive function.[2,3] Playing golf doesn't require high levels of physical fitness, which is one possible reason for its popularity among older individuals. However, the golf swing is a complex movement pattern that puts various joints of the body under stress, and golf participation has been shown to be responsible for injuries to the lower back, wrist, elbow, and other joints of the older golfer, which can also lead to a high risk of injury recurrence.[2]

Because of this, the importance of proper conditioning for the senior golfer should not be underestimated. Continued participation in golf can be a very important form of exercise and social interaction for an older adult. Furthermore, research reviews[2,4] conclude that conditioning programs, in addition to regular rounds of golf, are highly recommended for all older players irrespective of their level of participation. Not only could the programs prevent injury, they also have the potential to improve performance. Such programs do not need to be elaborate—home-based exercises incorporating one's own bodyweight, weighted clubs, or elastic tubing resistance are very effective.

▶ **Driving improvement** *What is clear is that for the aging golfer regular physical activity, such as a round of golf twice a week, improves musculoskeletal, cardiovascular, and cognitive function.[1–3] Evidence reveals that three 90-minute golf-specific exercise classes per week for eight weeks—in addition to playing—improve golf performance.[4] For a group of senior golfers, with an average age of 71 years, regular exercises that developed their functional ability— balance, stability, and mobility—resulted in an average improvement of 4 mph (6.3 km/h) in driver swing speed. Converting this to distance, given a calm day, would mean an extra 11–16 yards (10–15 m) of carry distance off the tee, and probably more roll distance as well.*

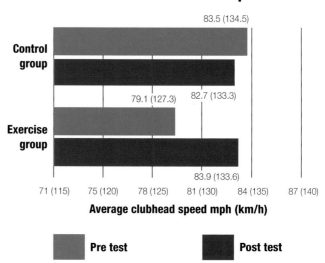

Increase in clubhead speed

Average clubhead speed mph (km/h)

- Control group: 83.5 (134.5)
- Exercise group: 79.1 (127.3) / 82.7 (133.3) / 83.9 (133.6)

Axis: 71 (115), 75 (120), 78 (125), 81 (130), 84 (135), 87 (140)

Pre test — Post test

Staying healthy

Brain
Cognitive decline is an evitable consequence of the aging process,[1] affecting abilities such as memory, reasoning and problem-solving. Long-term involvement in cognitively challenging activities, like playing golf, can result in changes to brain structure and function,[5,6] having a positive effect on cognitive activities.

Lungs
As we get older, the elasticity of our chest wall and lung tissue decreases, so we can't move air into and out of our lungs so efficiently. This means that, during physical activity, our breathing can't respond so well to provide the additional oxygen our tissues need, resulting in breathlessness. Regular exercise helps to improve and maintain lung function, so both the rate of breathing and the volume of air inhaled and exhaled in each breath increase more efficiently during exertion, and the sense of breathlessness is reduced.

Heart
Maximal aerobic function declines by appropriately 7% per decade from the age of 20, with cardiac function, expressed as maximal heart rate, decreasing by 4% per decade.[7] For the aging golfer, regular physical activity can maintain, and even improve, aerobic function.

Nervous system
The functioning of our nervous system declines with age. Nerve endings in the muscles, called proprioceptors, that send sensory information to the brain about the positioning of different parts of the body become less efficient, and the conduction velocity of electrical impulses along nerve axons slows. With regular activity older adults can help preserve neural function, and so maintain reaction times, balance, stability, and postural coordination.

Muscles (arm)
Depending on training and inherited traits, maximal muscle strength declines by an average of 1.5% per year after the age of 60, with the cross-sectional area of the muscle declining by 10% per decade after 50.[8] Maintaining physical activity levels plays a key role in preserving muscle mass and overall strength.

Fat deposits
A more sedentary aging lifestyle can lead to increased prevalence of abdominal fat deposition, with elevation in overall body fat and associated reductions in fat-free mass. Effects can include impaired mobility, increased risk of lower back pain, and greater threat of hypokinetic diseases. If an older person maintains an active lifestyle, however, these risks are greatly reduced.

Bone mass
Age-related declines in body mineral content, overall bone mass, cartilage water content, and joint lubrication increase vulnerability to injury and mechanical dysfunction. The gravitational loading and muscular traction that occur when walking around a course, improve bone thickness, strength, and calcium concentration.

Connective tissue
In older adults, connective tissue can become more stiff, brittle, and weak, with increased risk of tendon and ligament injury. It also has a decreased ability to return to its original length when injured, affecting stress properties. Physical activity is known to increase collagen turnover in connective tissue, which gives it improved pliability.

◀ **Peak condition** Golf presents both potential health benefits[2] and risks[3] for the senior player. The risks, such as musculoskeletal strains or cardiovascular stress, are compounded by the fact that the systems of older players may not be as efficient at withstanding the demands of this type of repetitive exercise. Research[2–4] has concluded that conditioning programs for the senior golfer are highly recommended, substantially improving health and golfing performance.

Does a balanced posture affect putting success?

How should I stand when putting?

Despite the relatively small body movements involved during the putting stroke, how a player stands and moves during those few brief seconds may reveal how posture at address and through the stroke could play a more important role than first thought in determining putting success. Top players look to create a stable, balanced, and solid base, along with a fixed pivot point to execute the stroke consistently.[1,2] Without these, the putting stroke may not stand up under pressure.

Using the latest scanning technology, a study published in the *Annual Review of Golf Coaching*[1] measured the pressure under the feet of both right-handed amateur and professional golfers while addressing the ball and making a strike. Recordings of the weight distribution between the right and left foot and the toes and heels revealed that amateurs place on average 20 percent more weight on the right side than left, with more pressure through their toes than heels. Professional players have a more even distribution, spreading pressure more consistently. When measuring the movement of pressure throughout the putt, the study also identified that amateur players created more sway during the putt while the professionals remained relatively still.

With uneven weight distribution at address, the researchers suggested that the amateur player is already placing their body in an "unbalanced" posture before they attempt the putt, meaning that any subsequent movements will simply be compensating. It has been found that many right-handed amateurs place a greater percentage of pressure on the right foot than the left, and more toward the right toe (vice versa for left-handed players) when standing still. When in the putting address posture the same pattern is observed; golfers with handicaps greater than 10 performed significantly worse than those with handicaps less than zero when both were asked to balance on one foot.[2] Given the importance of postural stability before and during the putting stroke, using activities that improve balance can lead to a more repeatable and mechanically sound ball strike.

▶ *Footwork A good putting technique creates a stable posture and pivot point to allow the putter to be returned consistently from address to impact without adjustment. Research suggests that variations in balance at address, and in the extent of pressure movement during the swing, may account for differences in putting success between amateurs and professionals.*

The right balance

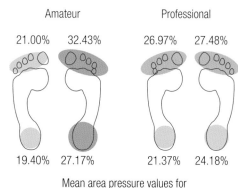

Amateur		Professional	
21.00%	32.43%	26.97%	27.48%
19.40%	27.17%	21.37%	24.18%

Mean area pressure values for different right-handed players

Pressure shift for amateur players

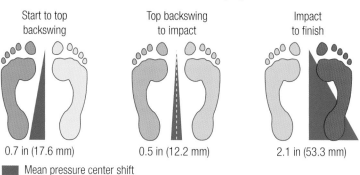

Start to top backswing	Top backswing to impact	Impact to finish
0.7 in (17.6 mm)	0.5 in (12.2 mm)	2.1 in (53.3 mm)

■ Mean pressure center shift

Pressure shift for professional players

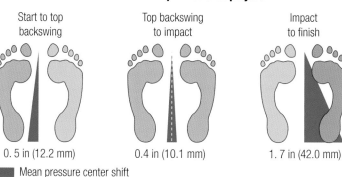

Start to top backswing	Top backswing to impact	Impact to finish
0. 5 in (12.2 mm)	0.4 in (10.1 mm)	1. 7 in (42.0 mm)

■ Mean pressure center shift

Pressure putting

The central nervous system, comprising the brain and spinal cord, processes the information received from the sense organs, coordinating the body.

The inner ears house the vestibular system, which monitors the directions of motion, such as moving forward-backward from toe to heel, or side-to-side from right to left.

The eyes observe where the body is in space and also the directions of motion.

Monitor display

Mechanoreceptors, positioned within muscles and joints, report what parts of the body are moving and where tension resides.

Pressure receptors located under the skin in the feet sense the distribution of pressure touching the ground.

Putting stance on pressure-sensing mat

▲ *Analyzing balance This illustrates a typical pressure image of a right-handed amateur golfer at the address position for a putt. The pale pink, dark pink, and red colors show the depth of pressure. For this amateur, as for many, the distribution of pressure is greater on the right foot and on the toes. Research[1] indicates that on average 60 percent is on the right foot and 40 percent on the left. Professional tournament players have a much more even distribution of pressure across left and right and heel and toe for putting (50 percent split).[1] Having an unequal balance before starting the stroke will certainly impact on the balance throughout the swing, and it is likely that this will begin unwanted movements to keep the club on line through impact.*

▲ *Putting posture An important component of the complete golf swing, good posture at the start and throughout the movement reflects good balance, stability, and mobility. Research shows that golfers with poor postural balance at address may lose rhythm or tempo, which affects their mechanical efficiency.[1]*

Does golf performance relate to perceived hole size?

Can imagining a bigger hole help me play better?

The history books reveal conflicting stories as to how the size of the hole was originally determined. Some sources suggest the hole size became standardized when golf officials began using a common drainage pipe to produce the hole. There is also reliable evidence that in 1829 officials at the Royal Musselburgh Golf Club in Scotland invented the first known hole-cutter. It produced a hole 4.25 in (10.8 cm) wide. From this point on, all holes on every course were standardized. However, even though actual hole size remains constant on modern courses, players' perceptions of dimensions on the green may vary considerably. A growing body of research demonstrates that how well the golfer is performing may actually affect how they perceive the size of a hole and distance to the pin. Evidence from the sport of softball has shown that players who are hitting well perceive the ball to be bigger than players who have more difficulty hitting.[1] Also, research shows that when study participants were asked to reach for a target with and without a tool (for example, a golf club), despite targets being presented at the same distances, perceived distances to targets within reach with the tool were compressed compared with those to targets that were beyond reach without it.[2] In another series of studies, golfers who played better, having lower scores on the course on that day, judged the size of a hole to be bigger than players who played worse,

having higher scores.[3] However, handicap, which is a measure of longer-term playing ability, did not significantly correlate with the judged hole size. Combined, these results suggest that better players did not see the hole as being bigger, but players that were playing better on that given day did. In a further experiment, players with a short putt of 1.3 ft (0.4 m) perceived the hole to be bigger than golfers with a longer 7 ft (2.1 m) putt. The results suggest that when players are faced with a shorter test they perceive the hole to be bigger, and success therefore more likely, than golfers facing a longer, more difficult putt. Since putting is harder from a farther distance and performance was markedly worse in the longer putt task relative to the shorter putt, these findings suggest that putting performance influences apparent hole size.

These studies suggest that a relationship exists between golf performance and perception of hole size, but the causal direction of the findings remains unclear. Do golfers putt better and therefore see the hole as bigger, or do they see the hole as bigger and therefore putt better? The research does not answer these questions, but it can be speculated that the relationship is reciprocal so that a golfer's perception of hole size and their putting performance are likely to influence each other.

▶ *Optical illusion* *More than any other aspect of golf, putting relies on one's eye—how one reads the undulating greens and finds the true path to the hole—and perception. To assess how the perception of the hole may both influence and reflect one's psychological state, Witt and colleagues' research recreated a well-known illusion: the smaller circles around the golf hole make the hole appear bigger than it really is, while the larger circles make the golf hole look smaller. This distortion of reality—called the Ebbinghaus illusion—has been well documented, and the researchers confirmed that the golfers' perceptions were indeed distorted as predicted.[4]*

Sizing up the hole

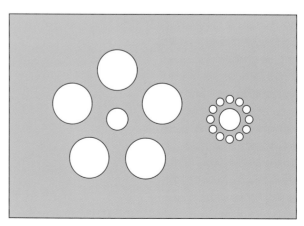

Size is everything

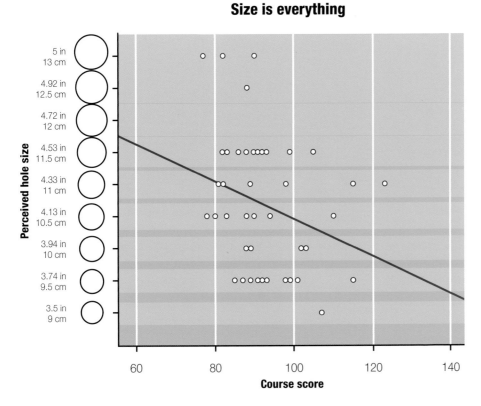

Perceived hole size (y-axis)

- 5 in / 13 cm
- 4.92 in / 12.5 cm
- 4.72 in / 12 cm
- 4.53 in / 11.5 cm
- 4.33 in / 11 cm
- 4.13 in / 10.5 cm
- 3.94 in / 10 cm
- 3.74 in / 9.5 cm
- 3.5 in / 9 cm

Course score (x-axis): 60, 80, 100, 120, 140

◀ ***Perception against performance*** *Evidence reveals that how a player perceives the size of a golf hole is a function of score on the course that day. In one such study[3] researchers asked players, once they'd finished their round, to select a hole size that best corresponded with the actual size of a golf hole (4.24 in/10.8 cm). With hole size increasing by a non-uniform diameter, a significant negative relationship (red line) was found between course score and the circle selected as best matching the actual size of the golf hole. Demonstrating that a golfer's perception of hole size is scaled by their current abilities, golfers who are playing better see the hole as bigger than do golfers who are not playing as well.*

Mind the hole

◀ ***Believing is seeing*** *Research indicates that the perception of the spatial layout of a golf hole may not only be a function of optical information about the cup, but also a function of the golfer's perception of their own immediate performance. In a series of experiments, scientists demonstrated that golfers perceive the size of a hole relative to their current golfing performance. In other words, players who played better, or putted closer to the cup, judged the hole to be bigger.*

equipment: golf shoe

The outcome of a golf swing is dependent upon precision, balance, and consistency. Every great swing has a starting point or setup, which places the golfer in an optimal position to execute a repeatable, efficient golf swing and ball impact. To generate the athletic movement of the swing, the relatively relaxed and stable posture must start from the ground up. After all, it is the interface between the ground and the foot that creates the stabilization upon which the swing movement can build and develop. During the initiation of the movement, the body builds muscular tension from the ground up, through the connection of the lower and then upper body segments, triggering the start of the backswing.

To ensure a stable platform for the initial accelerating physical forces of the swing, the golfer's connection with the ground needs to be optimized and, to this end, the modern golf shoe has evolved from humble origins about 150 years ago, as merely a boot with nails in the sole, into a high-tech golfing tool—more like an athlete's shoe.[1,2] Engineered to assist the player in making the push-off needed on power shots, the latest innovative designs allow the shoe to bend and twist while offering ample ground-hugging traction that helps stabilize the feet during powerful shots requiring significant leg push.

1920s

1940s

1950s

1960s

1856	1891	1906	1917	1923	1940
The earliest reference to spiked golf shoes appears in an issue of The Golfer's Manual. In this Scottish publication, novice golfers are advised to wear stout shoes "roughed with small nails or sprigs" to walk safely over slippery ground.	To provide more stability to the golfer, screw-in spikes were introduced. While they were more comfortable than the hob-nail shoes and boots worn by some golfers, during the next century groundskeepers complained about the spikes damaging the greens.	The first standard-looking golf shoe was introduced, which offered the golfer an extra saddle-shaped piece of leather around the laces. Named Saddle Oxfords, these were originally designed for racket sports, but gained more popularity among the golfing fraternity.	The first US patent was granted to William Park for his then "innovative" golf shoe design, which was more of a boot than a shoe! The materials used at this time were leather, wood, and canvas.	By the early twentieth century, the golf boot had turned into the golf shoe when the Field and Flint Company decided to improve men's golfing footwear.	It was during this year that women's shoes were first introduced in America. During these times the shoes looked like men's shoes.

Today's shoe

Incorporating a gel-based foot-bed with a reinforced heel support system, the modern shoe is both comfortable and functional, preventing unwanted movement and giving a firmer hold.

Decoupling grooves are incorporated within the sole to allow specific zones to move freely, keeping the foot more balanced and stable throughout the swing.

The evolution of the spike now offers superior traction and greater stability throughout all positions of the golf swing.

Optimized cushioning and improved energy return are provided by optimally placed shock absorption units throughout the sole.

Upper sole components of the shoe, which surround and bolster the foot, are strategically located to promote more lateral stability of the foot.

Innovation in the lacing system ensures the foot is kept stable and firm as forces act on it throughout the swing.

Thinner, more durable materials reduce heat retention, resulting in less fatigue during the round.

Lower profile outsole technology attempts to bring the foot even closer to the ground, offering improved stability, balance, and feel.

A reinforced toecap not only offers durability, but may also act to reduce front foot compression loads during the follow-through.

▶ **Stepping up** Golf shoes serve as a functional high-tech piece of equipment, offered with a variety of materials, soles, and cleats. The traction, flexion, moisture control, and comfort of golf shoes have been designed and engineered to deliver a state-of-the-art golfing tool.

The latest footwear is now built on a wider footplate, providing larger contact areas between shoe and turf. It's been suggested that this improves balance and stability by increasing connectivity throughout the swing.[1,2]

1970s	1980s	1990s	1995	2010
Advancements in manufacturing techniques and shoe design saw the first rubber-soled golf shoe appear on the market. This reduced shoe weight and improved balance and stability.	Waterproof-treated leather started to appear as a material of choice by major manufacturers. The shoe also started to become more athletic looking, resembling a sports shoe, focused on foot support and comfort.	The introduction of non-metal cleats on golf shoes gained popularity, offering better traction during the swing and causing less damage to the golf course.	More advanced materials offered more breathable yet fully waterproof uppers.	Golf shoe technology and design started to mimic advanced athletic shoes, allowing for a more natural motion of the foot, a substantial reduction in weight, and lower profile soles to improve balance and stability.

What does "functionally connected" mean?

→ Why are mobility and stability important?

Many coaches regard staying "connected" throughout a golf swing as maintaining a tight relationship between the body, arms, and club from start to finish. More scientifically, the concept of connectedness has both a mental and a physical dimension. From a psychological stance, being internally connected could be viewed as having an inner calmness, being present in the moment, and being relaxed and ready. Conversely, a state of disconnectedness is associated with anxiety, feelings of negativity, or self-doubt.[1]

From a biomechanical perspective, the notion of a connected swing can signify that all body segments are either accelerating or decelerating in the correct sequence with precise and specific timing so that the club arrives at impact accurately and with maximum speed.[2] Additionally, efficient connectivity between the body's neural pathways ensures communication between regions of the brain responsible for processes such as motor planning, control, estimation, prediction, and learning.[3]

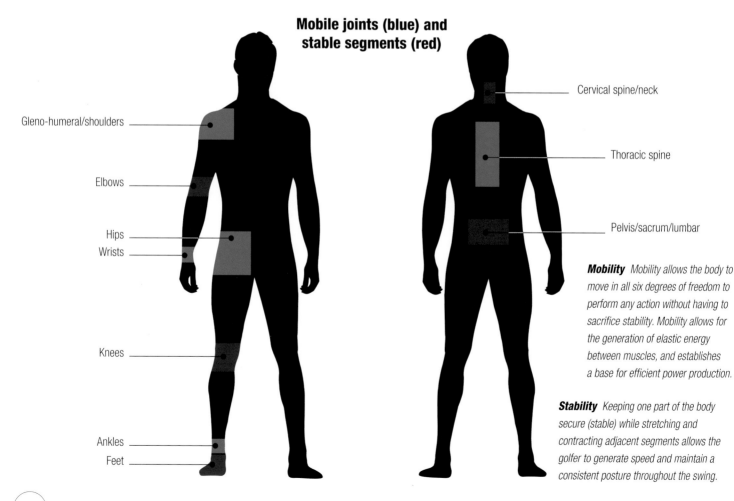

Mobile joints (blue) and stable segments (red)

Gleno-humeral/shoulders

Elbows

Hips

Wrists

Knees

Ankles

Feet

Cervical spine/neck

Thoracic spine

Pelvis/sacrum/lumbar

Mobility *Mobility allows the body to move in all six degrees of freedom to perform any action without having to sacrifice stability. Mobility allows for the generation of elastic energy between muscles, and establishes a base for efficient power production.*

Stability *Keeping one part of the body secure (stable) while stretching and contracting adjacent segments allows the golfer to generate speed and maintain a consistent posture throughout the swing.*

More broadly speaking, connectedness can refer to the ability to complete the desired pattern of movement, activating and steadying the correct body segments.[4] In turn, this will produce a consistent, powerful, and synchronized series of muscular activation patterns necessary for an efficient movement. If this series is altered, through poor technique, inflexibility, or muscular weaknesses, dysfunction will occur and the body will try to compensate, creating new and inconsistent body movements. So, from a physiological point of view, golfers need to ensure their body performs two important activities to stay functionally connected throughout their swing—effective regional mobility and stability. Mobility refers to the combination of muscle flexibility and joint range of motion, while stability is the ability to control the motion of a joint. During the golf swing, the body is an alternating pattern of stable segments and mobile joints. If there is a limitation in the mobility of a mobile joint, the stable region above or below will compensate to create movement and the stable segment will become mobile. When this sequence of muscle events is affected, the golfer becomes functionally disconnected, and the physical sequencing of the swing changes, in turn changing the swing's consistency, power, and accuracy.[2]

▼ **Mobility versus stability** *In order to create an efficient action, the body must operate in an alternating pattern of mobile joints and stable body segments. If this sequence is altered, dysfunction in movement patterns will occur. For the golf swing, this means that upsetting the combination of mobility and stability will adversely affect the execution of the swing, the body speed and transfer of this speed to the golf club.*

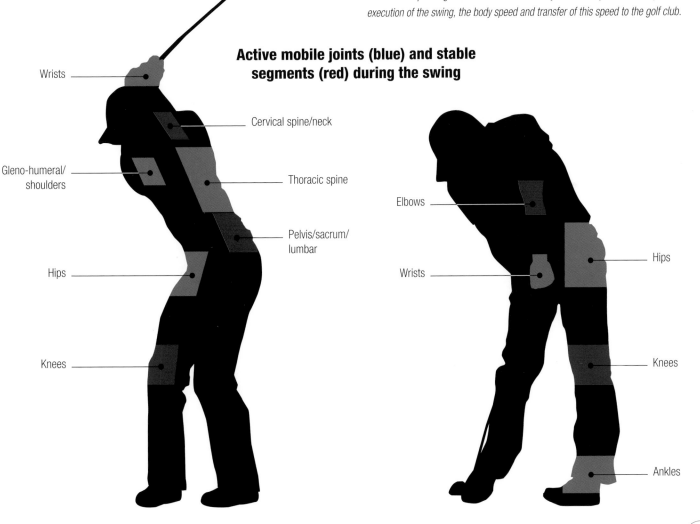

Active mobile joints (blue) and stable segments (red) during the swing

Wrists

Gleno-humeral/ shoulders

Hips

Knees

Cervical spine/neck

Thoracic spine

Pelvis/sacrum/ lumbar

Elbows

Wrists

Hips

Knees

Ankles

What can we learn from the brain activation patterns of top players?

What happens in the brain during the pre-shot routine?

Successful golfers have the canny ability to refocus after unwanted distractions, have control over their thoughts and emotions, and employ cognitive techniques in imagining intended shot outcomes. In addition to these characteristics, top performers can also deploy consistent cognitive-behavioral strategies that are maintained throughout competition. One specific cognitive-behavioral approach used in golf is the pre-shot routine. This ritualistic sequence of events that, time after time, prepares the golfer for their shot, is a process of mental and physical rehearsal. What a golfer thinks during these vital seconds before the swing may begin to reveal what is actually happening in the brains of the top players.

By using functional magnetic resonance imaging techniques, otherwise known as fMRI, scientists from across the globe are beginning to observe fairly striking differences in the brain activation patterns of players during those all-important seconds before the shot.[1-4] By examining the neural events in the brains of golfers while they are visualizing their normal golf swing,[2] or performing their mental pre-shot ritual,[3] researchers have begun to show that during these periods of mental rehearsal there is less neural activity in the brains of better players. At the lower-skill level, the typical swing is a complicated array of moves and adjustments, errors and corrections, anticipation and worry. Simply trying to organize thoughts and plan movements ahead of the strike can result in intense brain activity. With diminished brain activation occurring as skill level increases, it has been concluded that as a consequence of practice and experience, the tour player's brain becomes less activated during these periods as their movement creation and shot planning becomes more automatic.

▼ *Thought cycle Neuroimaging research has uncovered differences in the brain activation patterns of professional and amateur golfers while undertaking their normal pre-shot ritual.[2] Brain scans have revealed that areas of the inner brain, linked to functions such as emotional control, working memory, and topographical recall, seem more activated in amateurs than in professionals. It is thought that lower-skilled golfers—with less natural and automated movement patterns than professional golfers—make more conscious choices, allow distractions to flood the mind, and are unable to eliminate unwanted emotional thought. This in turn fires higher levels of brain activation.*

Control center

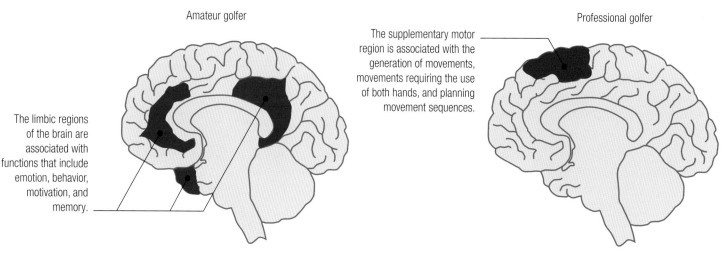

Amateur golfer

The limbic regions of the brain are associated with functions that include emotion, behavior, motivation, and memory.

The supplementary motor region is associated with the generation of movements, movements requiring the use of both hands, and planning movement sequences.

Professional golfer

Preparing for the shot

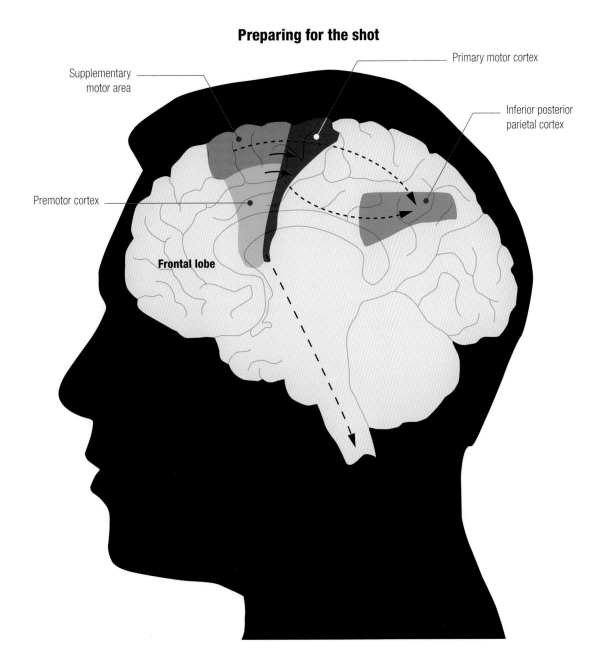

Supplementary motor area

Primary motor cortex

Inferior posterior parietal cortex

Premotor cortex

Frontal lobe

Although they represent very early stages of neuroscientific exploration inside the golfer's brain, these findings may just start to open a window into the inner workings of great players' minds. There still remain many unanswered questions, however. Would different fMRI patterns be seen in association with different clubs? Does the activation vary if the golfer imagines hitting over a huge hazard versus hitting down a broad, expansive fairway? Can this technique be used to improve play? Emerging neuroscience research may soon provide answers.

▲ *Pre-shot process* *So what may be happening in the brain of a Tour player when preparing to make that all-important shot? The creation of the golf swing movement during the pre-shot phase occurs within the premotor cortex, which prepares commands for voluntary actions triggered by the environmental context (such as distance to target, ground conditions, wind speed). The supplementary motor area prepares commands for internally generated "intentional" actions, which are then executed by the primary motor cortex. Signals containing copies of prepared motor commands (i.e. the swing) are also sent to the parietal cortex, where they are used to predict the sensory consequences of movement—in other words, to provide a mental picture of what the swing and shot outcome may look and feel like.*

What are the main mechanical stresses on the body during golf?

→ Will I get a bad back from playing golf?

The golf swing may appear to be relatively low-impact, given that the golfer remains in contact with the ground at all times, isn't hit by anything, and doesn't hit anything large. However, the explosive twisting, pulling, pushing, compressing, and bending motion during the swing causes considerable stress, particularly on the spine, which must withstand rotational loads caused when the upper body twists around the lower body. This is known as thorax–pelvis separation or the "X-factor stretch" (see pages 44–45), which occurs at the top of the backswing and the start of the downswing. The average recreational player, who is likely to play twice a week, swings the club nearly 20,000 times per year—can the body handle the mechanical loads placed on it when repeating this dynamic and explosive movement?

With the swing lasting little more than a second, and generating a clubhead speed of nearly 100 mph (160 km/h), the back has been calculated to produce compression loads in excess of 7000 newtons;[2] this means that for a 175 lb (80 kg) golfer there is the equivalent of around 10 times their own bodyweight acting on their spine. The back has evolved anatomically to provide reinforcement during compression, lateral bending, and torsion of the spinal disks to withstand these stresses during movement. However, amateur golfers can generate around 80 percent greater torque and shear loads than professional golfers because of an inefficient swing, thus leaving the weekend golfer prone to back problems.[2–4]

These problems may include muscle strains, slipped disk, or stress fractures of the vertebral body.[2] Most injuries of the back are cumulative—known as "cumulative trauma disorders" (CTDs).[8] A player who is out of shape, has poor address posture, or lacks mobility in the hips, mid-back, and shoulders, may eventually injure their back. The best way to improve your swing efficiency is to seek expert advice from a professional coach, who will also be able to advise on exercises to improve your mobility, flexibility, stability, and strength.

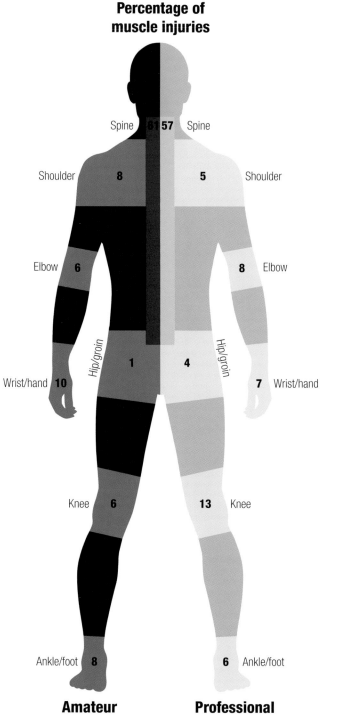

Percentage of muscle injuries

	Amateur	Professional
Spine	61	57
Shoulder	8	5
Elbow	6	8
Hip/groin	1	4
Wrist/hand	10	7
Knee	6	13
Ankle/foot	8	6

Amateur　　**Professional**

Twist and swing

The cervical or neck region of the spine consists of seven vertebrae, known as C1 to C7. The top cervical vertebra is connected to the base of the skull.

The thoracic region of the spine is located at the chest level, between the cervical and lumbar vertebrae. The 12 thoracic vertebrae, known as T1 to T12, also serve as attachments for the rib cage.

The lumbar region of the spine is located between the thoracic vertebrae and the sacrum. The five lumbar vertebrae, known as L1 to L5, are the main weight-bearing section of the spinal column.

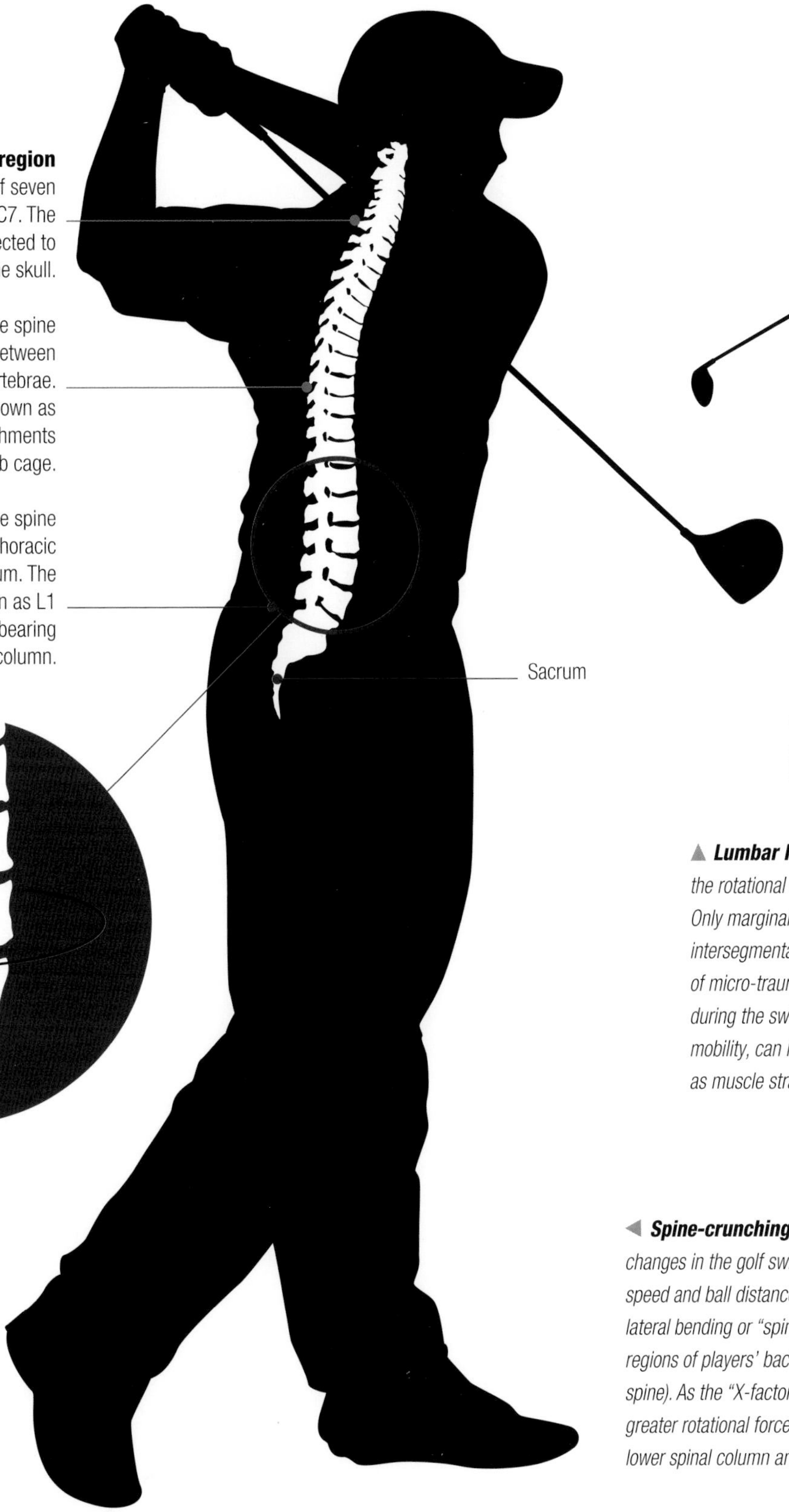

Sacrum

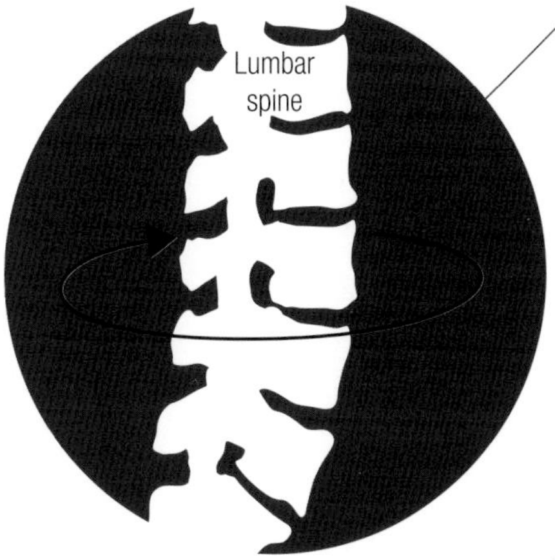

Lumbar spine

▲ *Torque talk* *The twisting force that occurs in the lumbar region during the swing may cause injury over time, if the golfer has not prepared and exercised adequately.*

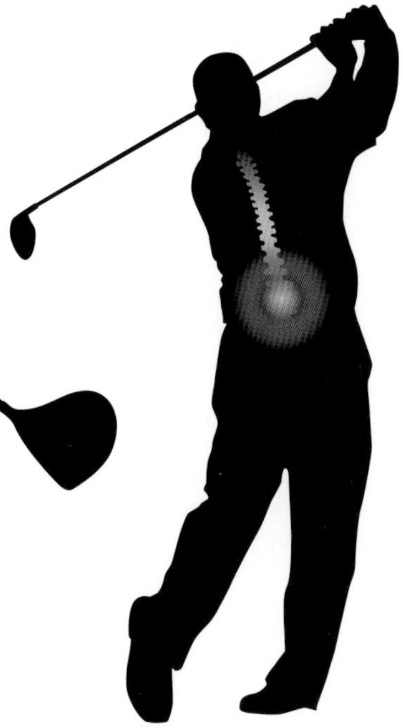

▲ *Lumbar loading* *The lumbar's key role during the rotational movement of the swing is stabilization. Only marginal rotation—between 2 and 3 degrees of intersegmental twist—may be enough to cause the onset of micro-trauma. Torsional loading of the lumbar spine during the swing, often caused by poor hip or mid-back mobility, can lead to acute local soft-tissue damage, such as muscle strain or internal disk disruption.* [2–4]

◄ *Spine-crunching* *It is agreed by many experts that changes in the golf swing over time—to emphasize swing speed and ball distance—have led to an exaggerated lateral bending or "spinal crunching" within the lower regions of players' backs (the backward "C" shape of the spine). As the "X-factor stretch" has increased to generate greater rotational forces, greater loads are placed on the lower spinal column and supporting muscular structures.*

remaining calm and confident

Anyone who plays golf regularly knows that learning on the driving range is one thing, putting all that into practice effectively and consistently on the course is something else. Maintaining a state of calm before, during, and after each shot, and retaining confidence throughout the round, are as crucial to a low score as a player's skill level.

No matter how skilled the player, a slip of concentration, a loss of focus, or dwelling on poor shots, all provide a recipe for disaster, but the physical state of a player also contributes to their mental state. Combining physical and mental discipline has been proven to help a player maintain their levels of concentration and so improve their performance on the course.

Good levels of physical fitness are the starting point for a winning mental strategy. Staying alert throughout all 18 holes requires stamina, and stronger muscles reduce the strain and tension a player may experience during each shot. Physical fatigue will begin to reduce the effectiveness of the body and the brain later in the round. This also means that a player needs to take on the right levels of nutrition and fluids to maintain physical performance and mental acuity. Leading players also learn to control their breathing and their heart rate before and during a shot. This helps their muscles to relax and execute the movements that they've been trained to perform during practice.

Tour pros also use various techniques, such as the pre-shot and "quiet-eye" routines, to reduce the neural activity in their brains, eliminate distractions, reduce cognitive anxiety, and help them focus on making the shot. A number of studies[1,2] have indicated that the amount of cognitive anxiety—in other words, doubts about performance before and during play—is less important to actual performance than the direction of that anxiety, that is, whether the player uses that anxiety to be ultimately helpful to themselves.

▶ **Staying in the present** *The course is full of hazards waiting to penalize any golfer for wayward shots. Keeping calm and composed, even when your game plan falters, requires confidence, focus, and control of negative emotions. Having placed his tee shot into the water, Tiger Woods stays calm and takes a drop on the par-3 15th during his first round of the 93rd PGA Championships.*

Do hydration levels affect mental and physical performance?

→ **Will drinking water regularly help my score?**

Playing golf—particularly in warm weather—leads to sweat loss and dehydration, and this can have an impact on both mental and physical functions during recreational and competitive play. Dehydration has been shown to reduce motor performance, cognitive function, and alertness in a range of athletic and non-athletic groups. Research using cognitive-motor tasks to measure perceptual discrimination, target accuracy, visual tracking, choice reaction time, attentional focus, concentration, and fatigue perception has concluded that the effects of mild dehydration—of 1–3 percent ΔBM (change in body mass)—result in cognitive-motor dysfunction. Whether such mild dehydration impairs neurophysiological function during golf-specific cognitive-motor performance has yet to be fully explored, but new research has started to reveal that not replenishing lost fluids while out on a course can actually result in an increase in the number of errors a player makes, affecting their score.

Findings from a recent study published in the *Journal of Strength and Conditioning Research* support previous suggestions that mild dehydration—a reduction of around 1–2 percent in body mass—significantly impairs mental and physical function during golf.[3] This study was the first to show that mild dehydration can adversely affect hitting distance, accuracy, and judgment of distance during play. The results demonstrated that even a small reduction in body mass,

attributed to acute mild dehydration (a mean of −1.5 percent ΔBM), augmented by an absence of fluid intake for 12 hours, impairs golf performance. Given that players are on-course for periods of around four hours, often with limited opportunities to take on fluids, maintaining hydration levels throughout a round can be compromised.

Currently, researchers still acknowledge that there is no consensus as to whether reduced cognitive-motor function, increased by mild dehydration, is a consequence of physiological homeostatic imbalance—such as hormonal or cellular effects resulting from dehydration—or caused by central motor behavior changes attributable to increased sensations, such as thirst. New evidence appears to indicate that impaired cognitive-motor performance may actually be linked to a centrally mediated mechanism of thirst perception, rather than to a loss of total body water. In other words, a signaling mechanism in the body may promote a greater conscious perception of effort in order to encourage a change in behavior.

▶ **Water** *As the single largest component of the human body, water accounts for around 60 percent of the total body mass. For a healthy golfer with a body mass of around 155 lb (70 kg), that makes up about 11 gallons (42 liters). The performance of prolonged exercise, like golf, particularly in warm environments, can result in a substantial loss of body water, with the potential for adverse effects on performance capacity.*

Water as percentage of body mass

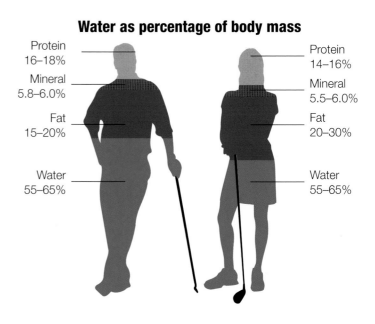

Protein
16–18%

Mineral
5.8–6.0%

Fat
15–20%

Water
55–65%

Protein
14–16%

Mineral
5.5–6.0%

Fat
20–30%

Water
55–65%

Effects of dehydration

Mental performance

1 Mental tiredness will increase and both alertness and concentration will be reduced.[1,4]

2 Golf skill and accuracy will be affected, with reduced coordination meaning the player will hit the target less often and with poorer precision.[3]

3 Decision-making will be impaired, impacting club and shot choice as well as judgment of distance.[3]

Physical performance

4 Cardiovascular and central nervous systems will be affected, causing increase in heart rate, lower blood pressure, and loss of muscle strength.

5 Physical fatigue will result in loss of coordination, balance, and stability, affecting shot accuracy and distance, and movements will appear to require more effort.[2,3]

6 Even a small reduction in body mass resulting from dehydration can reduce muscular strength by up to 6%.[1]

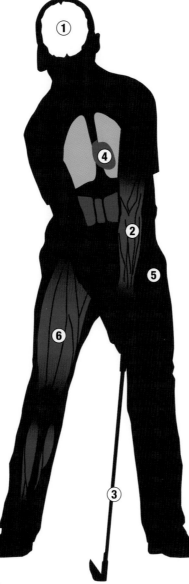

▶ **Water shortage** *According to the latest research, mild dehydration can affect mood, energy levels, and the ability to think clearly. Hydration experts say that our thirst sensation doesn't usually appear until there has been a reduction of 1–2 percent in body mass, when dehydration has already occurred. By this stage, both our mental and physical performance may have already felt the impact.*

Consequences of water loss

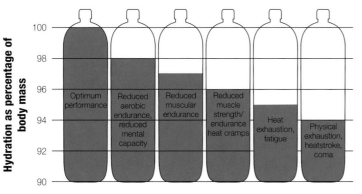

Hydration as percentage of body mass (y-axis: 90, 92, 94, 96, 98, 100)

- Optimum performance
- Reduced aerobic endurance, reduced mental capacity
- Reduced muscular endurance
- Reduced muscle strength/endurance heat cramps
- Heat exhaustion, fatigue
- Physical exhaustion, heatstroke, coma

◀ **Go with the flow** *A 200 lb (91 kg) player would only have to lose on average 3 lb or 1.4 kg (–1.5 percent △BM)—that is, 12 oz (350 ml) of fluid depletion per hour during a four-hour round of golf—to impact on mental and physical performance during the final, often crucial, holes. With possible sweat rates in temperate conditions (64–72ºF or 18–22ºC) of around 13.5 oz (400 ml) per hour when walking at around 3 mph (5 km/h)[1] a player must maintain fluid replacement during a round.*

Does cardio-respiratory response affect putting?

Should I breathe in or out to hole a putt?

It is a long-held belief, pioneered in the early 1970s by Professor Herbert Benson, that a relationship exists between the control of the body's physiology and the functioning of the mind during moments of stress. For a golfer under pressure, when the game is on a knife's edge, being able to regulate bodily responses by focusing attention in particular ways could mean the difference between holing that game-winning putt, or missing it.

In 2009, researchers from Australia stumbled across a theory known as the intake-rejection hypothesis that may in part explain why successful golfers are more likely to sink more game-winning putts than novice players.[1] Exploring the relationship between playing ability and key physiological responses during the putting stroke, it was found that those who were most successful at sinking an 8 ft (2.4 m) putt tended to demonstrate different cardiac and respiratory responses before and during the stroke. Within the complex neuro-circuitry of the body—influencing physiological, emotional, and cognitive processes—it would appear that these physiological differences may have been brought about by attentional processes within the brains of the higher-skilled golfers.

According to the theory, better players are more effective at encoding environmental information in order to execute the putt. Elite golfers have developed a higher level of automatic movement, and therefore are able to trust their swing and focus more attention on external goal-oriented factors, such as the target, ball, or putting line rather than how their body feels. The physiological consequence is that the heart rate slows, and the lungs deflate, momentarily before the all-important strike. However, when novices internalize their attentional focus, for example on the head or the grip, the heart rate accelerates and the lungs fill.[1,2] With elite golfers externalizing thoughts more than beginners,[3,4] we can begin to see how the integration of body and mind can improve the golfer's focus and be a determining factor in putting success.

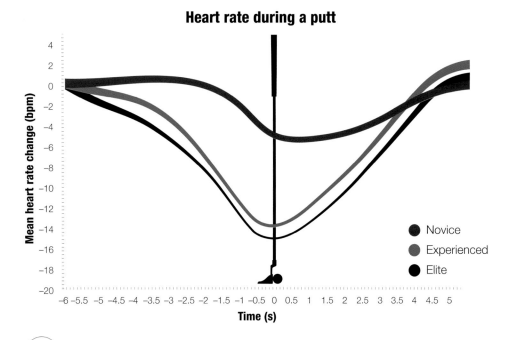

Heart rate during a putt

- ● Novice
- ● Experienced
- ● Elite

(y-axis: Mean heart rate change (bpm); x-axis: Time (s))

◀ *Mind over matter Ensuring current thoughts are in the right place before making a putt can have dramatic effects on physiology. As shown in the graph, experienced golfers are able to lower heart rate response to a greater extent moments before making contact with the ball. This reflects an ability to construct a pre-shot "mind routine" that focuses on environmental cues linked to the task and not internal cues from the body.*

Control of heart rate

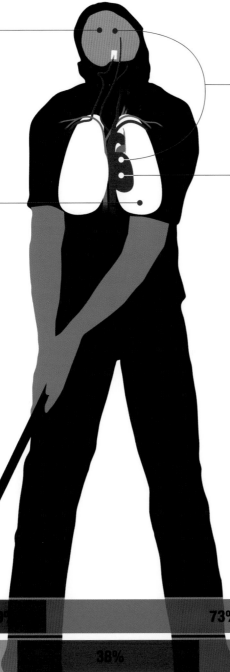

During the putt, the prefrontal regions of the brain interpret task demands while the sensorimotor regions receive information about the body, creating a motor command to execute the stroke.

Inhalation of air into the lungs leads to an increase in heart rate known as respiratory sinus arrhythmia, while breathing out increases parasympathetic nerve activity, lowering the heart rate.

Higher-order brain regions affect the control of heart rate by influencing the neural inputs to the heart, maintaining its muscle tone, and altering its pace.

Heart rate is controlled by the competing sympathetic and parasympathetic nervous systems, which increase and decrease the rate, respectively. What happens in our brain can profoundly influence the heart rate response, reflected through the variability between beats.

◄ **Brain and breath** *The execution of a psychophysiologically complex movement like a golf putt requires a neural superhighway of connections sending and receiving signals to and from the brain. The neural pathways between the cardio-respiratory "machinery" and the brain "mainframe" are so interconnected that subtle changes in cognitive-attentional processes can have profound effects on heart rate and breathing response. Research has shown that training attentional focus and breathing should lead to more putts being holed.[1–2]*

▼ **Let it out** *The graph depicts the breathing patterns immediately prior to an 8 ft (2.4 m) putt in three different golfing abilities. Elite golfers are much more likely to exhale just before making contact with the ball, releasing muscle tension, removing unwanted thoughts, and allowing focus on the external aspects of the putt.*

Inhale
Hold
Exhale

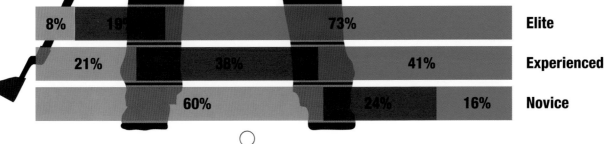

8%	19%	73%	**Elite**
21%	38%	41%	**Experienced**
60%	24%	16%	**Novice**

Of all the techniques used to play sport, the golf swing has perhaps been scrutinized more than any other. There are now literally hundreds of instructional books describing the golf swing, but many of these are based on anecdotal insights provided by elite players and their coaches. In this chapter, various aspects of the golf swing are considered from an evidence-based scientific perspective. Integrating theoretical and empirical research from the sub-disciplines of sports biomechanics and motor control, Paul Glazier and Peter Lamb tackle a series of important questions about the sequencing and timing of body segment rotations, the magnitude and variability of grip forces, and the transference of weight during the swing, among others.

chapter two

the swing

Paul Glazier and Peter Lamb

What is the "summation of speed" principle?

→ ## How can I coordinate my movements to maximize clubhead speed?

In recent years, the sequencing and timing of body segment motions in the golf swing have been popular topics for study by golf scientists, in part because of the more widespread use of automated 3D motion-capture technology. An important finding to emerge from this line of research is that the generation of clubhead speed in golf is governed by the "summation of speed" principle.[1] This principle states that to maximize velocity at the end of a series of linked body segments (known as a "kinematic chain"), the series should commence with the larger, heavier, inner (or proximal) body segments and proceed to the smaller, lighter, outer (or distal) body segments.[2]

A number of biomechanical studies examining proximal-to-distal sequencing in the golf swing have generally confirmed that, following the preparatory (backswing) phase, the action (downswing) phase starts with a rapid rotation of the pelvis (498 degree/s), followed by progressively faster rotations of the thorax (723 degree/s), lead arm (1165 degree/s), and club (2090 degree/s) as impact nears, before slowing during the recovery (follow-through) phase (representative data taken from Geisler[3] for professional golfers).

Another study, by Cheetham and colleagues,[4] has compared the kinematic chains of professional and amateur golfers. It was reported that the professional golfers exhibited significantly greater values than amateurs for the following variables: all average rotational accelerations and decelerations (except pelvis); all peak rotational speeds; all rotational speed gains (i.e. pelvis to thorax, thorax to arm, and arm to club); and peak linear clubhead speed.[4] It was suggested that, in general, these results indicated that amateurs had poorer coordination, weaker power production, and less efficient energy transfer between segments than professional golfers.

Based on these findings, it would appear that analyzing the motions of body segments in the kinematic chain could provide invaluable information for identifying faults and prescribing modifications to technique. With the increased availability and affordability of 3D motion-capture technology and its growing capacity to more accurately measure body segment motion in realtime, this line of research could help professionals and amateurs alike to improve the efficiency of their swings and maximize clubhead speed.

Kinematic chains

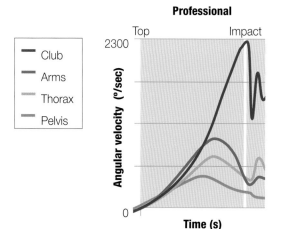

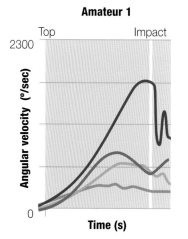

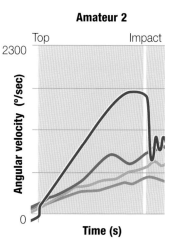

Accelerate to accumulate

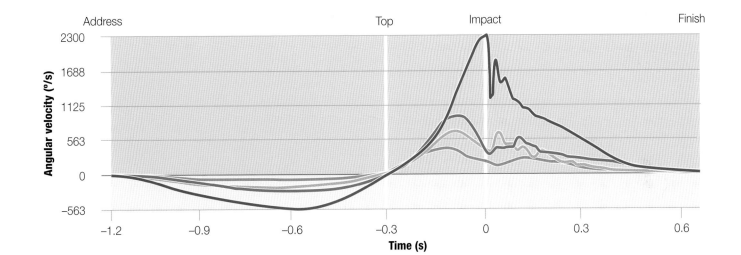

◀ **Smooth swing** The kinematic chains for one professional and two amateur golfers during the downswing phase (adapted from Cheetham et al[1]). The professional exhibits smooth accelerations and decelerations, with each curve peaking higher and later than the previous one, and the club peaking at impact. Amateur 1 exhibits poorer accelerations and decelerations and lower angular velocities, and the arm peaks before the thorax. Amateur 2 exhibits no overall decelerations of the pelvis and thorax before impact.

— Club
— Arms
— Thorax
— Pelvis

▲ **Phase change** The graph shows changes in angular velocities of body segments in the kinematic chain during different phases of the golf swing. During the downswing phase (period between the top of the backswing and impact) there is a sequential increase in angular velocity from the most proximal segment (pelvis) to the most distal segment (arm) and golf club.

Does head movement hinder swing performance?

Should I keep my head still during my swing?

Head movement is a good marker for body movement in the golf swing, specifically the lateral thorax shift. Although thorax anterior–posterior (or forward–backward) tilt can be compensated for by knee and hip flexion–extension, lateral thorax shift typically coincides with lateral head movement. This may be the reason that coaches have long viewed head movement as having such importance. The "head still dogma" was supported by the findings of Cochran and Stobbs,[1] who proposed a "pendulum" model with a fixed pivot point, or fulcrum, located midway between the shoulders on the golfer. The idea was that keeping the axis of rotation immobile should result in a repeatable swinging motion of the pendulum. However, recent research has shown that the axis does not remain still during the full swing.[2] Simulations have also shown that immobilizing the fulcrum has a negative effect on clubhead speed.[2] In fact, to keep the head perfectly still during the swing requires the golfer to dissociate their head movement from the rest the body's moving parts. It may be easier for the brain to couple head movement with other moving body parts than to try to operate each independently.[3]

Horan and Kavanagh[4] have shown that head movement patterns for professional golfers are consistent but vary from individual to individual. A common assumption is that less-skilled golfers move their head too much compared with expert golfers, but Sanders and Owens[5] have reported that, for a small sample of expert and novice golfers (six of each), novices actually displayed less total head movement throughout the swing compared with expert golfers. The experts, however, showed more consistent head movement during late downswing and through impact, keeping their head behind the ball through impact more than the novices. Coaches may use head movement as an indicator of body movement (particularly lateral shift), but for the full swing there is no reason to advocate a particular pattern of head movement—including the *absence* of head movement.

▶ *Head masters* It has been shown[3] that—for putting and possibly also chipping—expert golfers move their head in an egocentric pattern (i.e. the clubhead moves away from the target, and the head moves toward the target) relative to the club as compared with beginners who move their head in an allocentric pattern (i.e. clubhead moves back, away from the target, and head also moves back). Expert golfers also felt that their head stayed still while performing the egocentric movement pattern, even though it didn't, although the range of movement was minimal. It is recommended that golfers avoid the allocentric head movement pattern typically adopted by beginners.

Eye movement

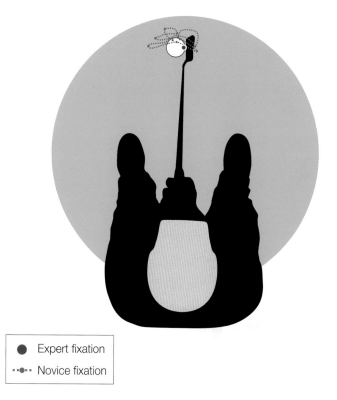

●	Expert fixation
⋯•⋯	Novice fixation

▲ *Eye on the ball* Mann and colleagues[6] showed that expert golfers fixate their focal gaze more consistently and for longer (red) on the ball, and on the target, compared with high-handicap golfers (blue). This phenomenon may give the feeling of a still head position during putting and may well be responsible for expert golfers reporting that their heads remained still during putting and in the full swing.

Head and putter movement

Allocentric pattern (generally used by novices)

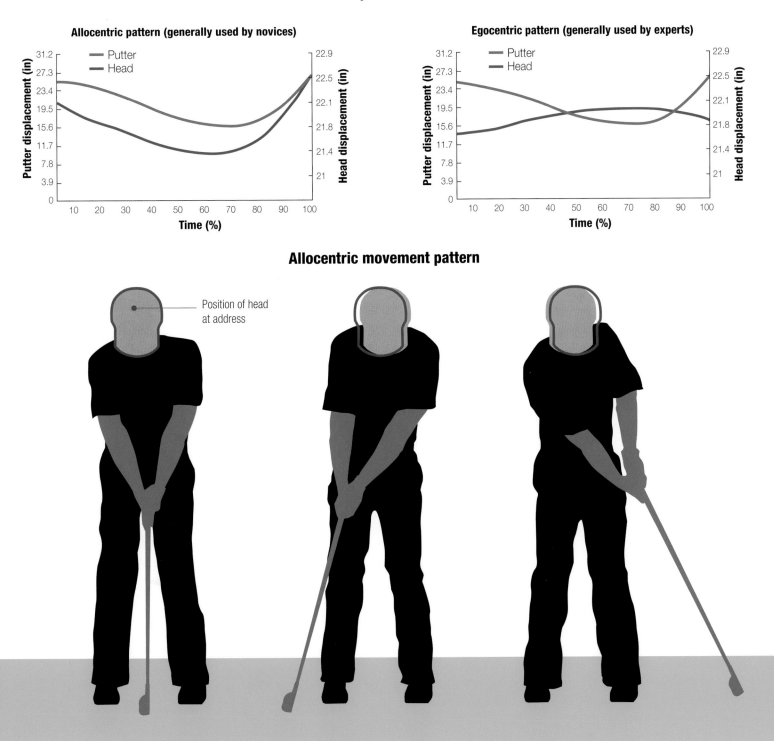

Putter displacement (in)
Head displacement (in)
Time (%)

— Putter
— Head

Egocentric pattern (generally used by experts)

Putter displacement (in)
Head displacement (in)
Time (%)

— Putter
— Head

Allocentric movement pattern

Position of head at address

What is the "X-factor" and what role does it play in hitting a golf ball far?

Does a greater body turn help me hit farther?

The "X-factor" is a buzzword often found in contemporary golf instruction manuals. It was first introduced by Jim McLean in a *Golf Magazine* article in 1992[1] and refers to the differential in degrees of rotation between the pelvis and thorax during the golf swing. Based on data subsequently published as part of a larger study by McTeigue and colleagues,[2] McLean showed how five of the longest-hitting professionals of the PGA Tour had a larger X-factor at the top of the backswing than five of the shortest-hitting professionals. He proposed that a direct relationship exists between the pelvis–thorax separation angle at the top of the backswing and clubhead speed at impact— that is, a larger separation angle is generally associated with greater clubhead speed, and longer drives. Many coaches now advocate a restricted pelvis turn during backswing to maximize the relative angle between the pelvis and thorax.

An increase in the X-factor during the early downswing has also been shown to be associated with higher clubhead speeds and longer distances. This move—called the "X-factor stretch" [3]— involves the rotation of the pelvis back toward the target while the thorax is still either completing the backswing or is stationary at the top of the backswing. It has been hypothesized that, by increasing the separation angle between the pelvis and thorax during transition, the abdominal and oblique muscles are dynamically stretched, leading to a more forcible contraction and a greater transfer of energy to the thorax. Empirical support for the role of the X-factor stretch was initially provided by Cheetham and colleagues,[3] who showed that a highly skilled group of golfers exhibited a significantly greater X-factor stretch than a lesser skilled group. Interestingly, they also found that no significant difference in the X-factor at the top of the backswing existed between groups. These findings led the researchers to suggest that the X-factor stretch is potentially more important than the X-factor when generating high clubhead speeds, although more recent research[4, 5] has indicated that they are of equal importance.

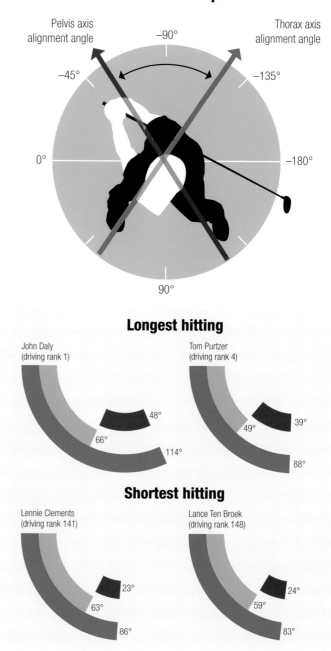

"X" marks the spot

Pelvis axis alignment angle

Thorax axis alignment angle

−90°

−45°

−135°

0°

−180°

90°

Longest hitting

John Daly (driving rank 1)

48°
66°
114°

Tom Purtzer (driving rank 4)

39°
49°
88°

Shortest hitting

Lennie Clements (driving rank 141)

23°
63°
86°

Lance Ten Broek (driving rank 148)

24°
59°
83°

X-factor differences

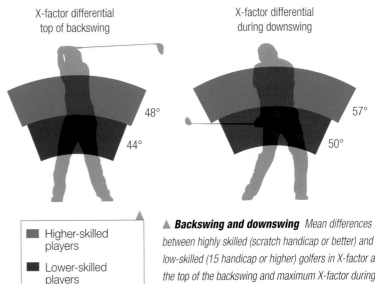

X-factor differential
top of backswing

48°
44°

X-factor differential
during downswing

57°
50°

Angle of separation *The "X-factor" is defined as the separation angle between the pelvis and thorax during the golf swing, specifically at the top of the backswing. When viewed from above, the intersection of the pelvis axis (defined as an imaginary straight line running through the hip joints—red line) and the thorax axis (defined as an imaginary straight line running through the shoulder joints—blue line) in the horizontal plane forms an "X," hence the term "X-factor."*

▼ **"X" statistics** *The longest- and the shortest-hitting golfers on the PGA Tour on average did not differ in the amount of thorax rotation (88° versus 89°) but they did differ in the amount of pelvis rotation (50° versus 65°).[1] When expressed as a percentage of thorax rotation, the differential between the thorax and pelvis rotations, or the X-factor, accounted for 43 percent and 27 percent for the longest and shortest hitters, respectively. Interestingly, John Daly—the number one ranked driver in 1992—had the third-largest pelvis rotation of all the tournament professionals analyzed. He also had the largest X-factor. These findings indicate that, contrary to popular belief, a restricted pelvis rotation is not a prerequisite for a large X-factor. These findings also suggest that caution should be applied when attempting to extrapolate results from group-based analyses to specific individuals.*

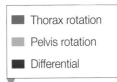

■ Higher-skilled players
■ Lower-skilled players

■ Thorax rotation
■ Pelvis rotation
■ Differential

▲ **Backswing and downswing** *Mean differences between highly skilled (scratch handicap or better) and low-skilled (15 handicap or higher) golfers in X-factor at the top of the backswing and maximum X-factor during the downswing, as reported by Cheetham and colleagues.[3] On average, the highly skilled golfers stretched their X-factor by 19 percent at the beginning of the downswing, which was significantly greater than the 13 percent stretch exhibited by low-skilled golfers. However, there was no statistically significant difference between the two groups in the X-factor at the top of the backswing.*

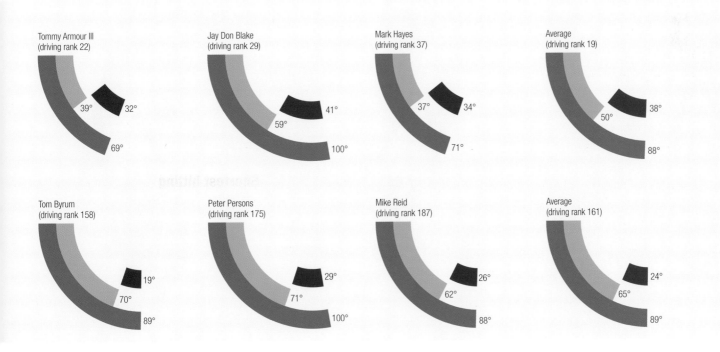

Tommy Armour III (driving rank 22)
39° 32° 69°

Jay Don Blake (driving rank 29)
41° 59° 100°

Mark Hayes (driving rank 37)
37° 34° 71°

Average (driving rank 19)
50° 38° 88°

Tom Byrum (driving rank 158)
19° 70° 89°

Peter Persons (driving rank 175)
29° 71° 100°

Mike Reid (driving rank 187)
26° 62° 88°

Average (driving rank 161)
24° 65° 89°

What neuromuscular patterns characterize an effective golf swing?

What muscles do I use during my golf swing?

Knowledge of the muscles that are most active at various times during the golf swing is important, not only from a golf coaching standpoint, but also from a strength and conditioning perspective. Using a technique known as electromyography (or EMG), scientists have been able to shed some light on how the muscles work during the golf swing. By attaching electrodes to the surface of the skin adjacent to specific muscle groups or by inserting fine-wire indwelling electrodes directly into individual muscles, the magnitude and duration of electrical activity within the muscle(s) can be measured.

Although it is impractical to analyze the action of all muscles simultaneously, a number of studies have considered the activation characteristics of key muscles of the trunk, shoulder, chest, forearm, and lower limbs during the golf swing.[1,2] When

combined, these provide a reasonable picture of the sequencing and timing of muscle activity during the golf swing. Most studies have examined muscle activity during the following phases: backswing; forward swing; acceleration phase; early follow-through; and late follow-through.

The information presented here may help to identify where musculotendinous injuries may occur, and to direct muscle strengthening programs that can increase the robustness of muscle and connective tissue to acute and chronic trauma. Further research is required on mid- to high-handicap golfers as most studies, to date, have focused almost exclusively on low-handicap or professional golfers. More studies investigating muscle activity in female golfers are necessary as much of the extant literature has considered only male golfers.

Muscle activity

▶ *Backswing*
The most active muscles during the backswing (the period between address and the top of the swing) are the right trapezius (upper, middle, and lower portions), right rhomboid, right levator scapulae, left serratus anterior (upper and lower portions), and left subscapularis. [It is worth noting that the situation is mirrored for left-handed golfers.]

▶ *Forward swing*
The most active muscles during the forward swing (between the top of the backswing and the horizontal position of the club during the downswing) are the right gluteus maximus (upper and lower portions), right gluteus medius, right biceps femoris, right semimembranosus, left adductor magnus, left biceps femoris, and left vastus lateralis.

Muscle map

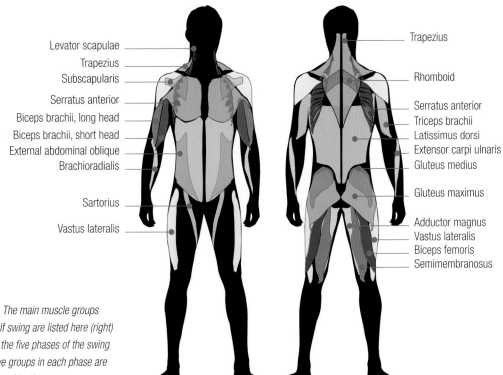

Levator scapulae
Trapezius
Subscapularis
Serratus anterior
Biceps brachii, long head
Biceps brachii, short head
External abdominal oblique
Brachioradialis

Sartorius

Vastus lateralis

Trapezius

Rhomboid

Serratus anterior
Triceps brachii
Latissimus dorsi
Extensor carpi ulnaris
Gluteus medius

Gluteus maximus

Adductor magnus
Vastus lateralis
Biceps femoris
Semimembranosus

▶ *Mighty muscles* *The main muscle groups recruited during the golf swing are listed here (right) and highlighted during the five phases of the swing (below). The more active groups in each phase are shown in progressively redder tints.*

▶ **Acceleration**
The acceleration phase is the movement between the horizontal position of the club during the downswing and ball impact. The most active muscles during this phase are the serratus anterior, external abdominal obliques, and left biceps femoris.

▶ **Early follow-through** *The early follow-through is defined as the period between ball impact and horizontal position of the club during the follow-through. The most active muscles during this phase are the serratus anterior and left biceps femoris.*

▶ **Late follow-through** *The late follow-through is defined as the period between horizontal position of the club during the follow-through and the end of the swing. Here the active muscles are latissimus dorsi, external abdominal obliques, bicep femoris, and gluteus maximus.*

47

equipment: the glove

Golf is one of very few sports that see the performer wear only one glove, instead of a pair. Propelled by a better understanding and application of materials science and technology, the glove has come a very long way from its original concept in the late 1800s as a simple means of protection against unwanted calluses. Popular magazines of the time promoted a glove specifically targeted at golfers, with pleats offering more room for movement around the knuckles. These first gloves were generally either fingerless or backless and ensured that the hands remained blister-free.

It wasn't until the 1930s, when professional golfers began to wear gloves, that their popularity grew. In addition to protecting the hand from the wear and blistering from repeated swings, the glove also offered a layer between the backhand and club, thereby creating a firmer grip throughout the swing. Eager to try anything that might improve their game, touring professionals quickly adopted the glove as standard equipment. As new players appeared, they were more likely to have already begun playing golf with a glove, and so, by the 1950s and 1960s, it had become firmly embedded within the golfer's bag. Modern breathable and waterproof materials, combined with new machining and sewing techniques, have led to today's high-tech piece of equipment.

Generating friction

◄ **Get a grip** The interaction between the glove, bare hand, and club grip provides forces of friction that ensure, whether in dry or wet conditions, that the hands don't slip relative to the club or to each other. At address, the outer surface of the glove and the grip remain in firm contact with each other, therefore locking together. This is known as static friction. As the club begins to move during the swing, the bare skin, inner glove surfaces, the glove's outer surfaces, and the grip can start to move if the club is moving relative to the hands. This gives rise to kinetic friction.

Mathematically, the maximum friction available between the glove and grip (F_f) is a function of the materials comprising the two surfaces (reflected in the coefficient of friction, μ) and the grip force, which is the normal or perpendicular force pushing the two objects together (F_n).

$$F_f = \mu F_n$$

If the glove (or grip) is worn or wet, the coefficient of friction will be reduced, making it more likely that the hands will slip.

Comfort and performance

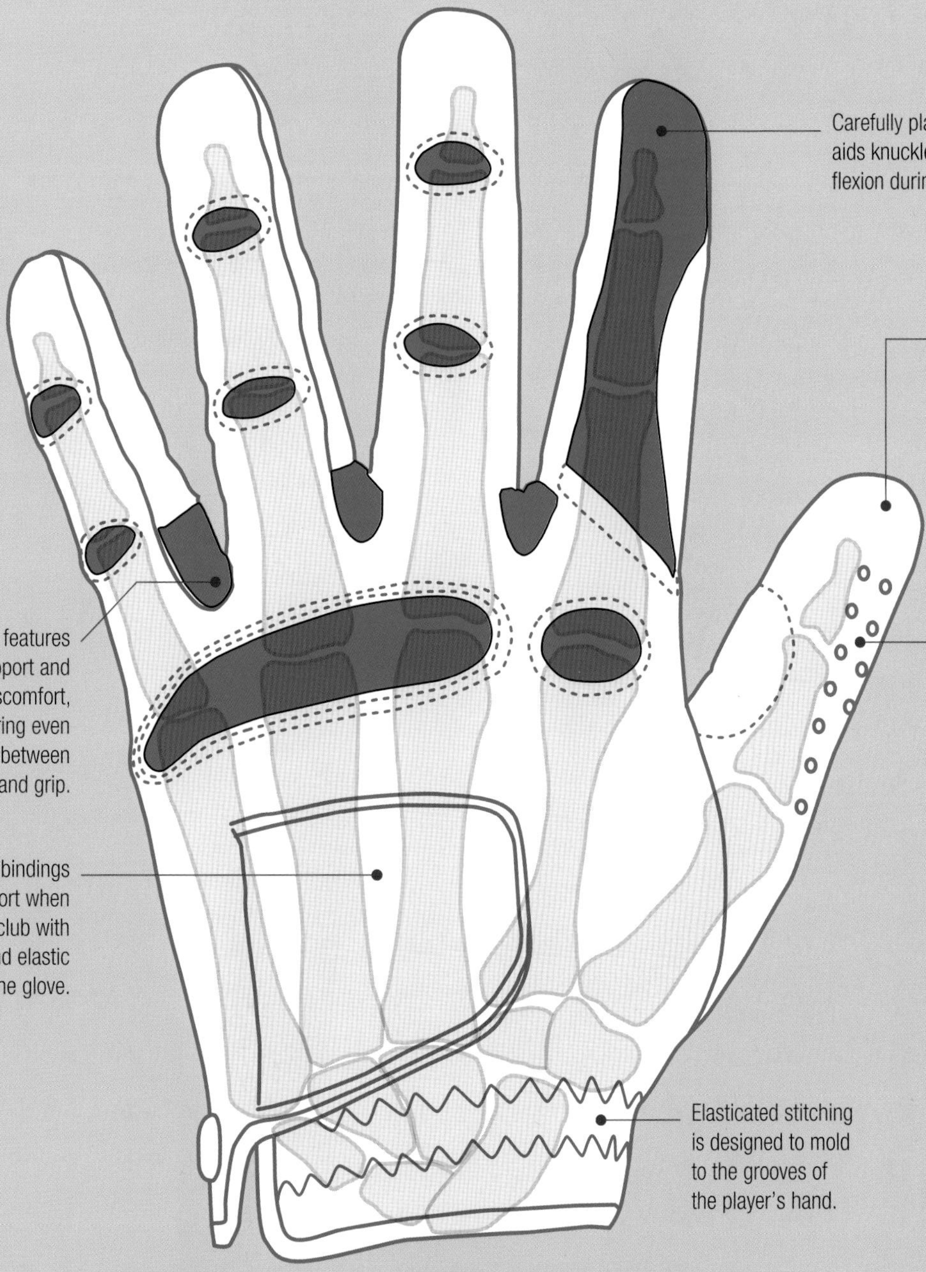

Carefully placed mesh aids knuckle and finger flexion during the swing.

Advanced fabric technology ensures the material of the glove remains soft to the touch and keeps its shape.

Ergonomic features provide support and eliminate discomfort, while ensuring even contact between glove and grip.

Modern breathable materials and ventilation holes keep the hand cool and reduce unwanted moisture.

Smooth bindings provide comfort when gripping the club with both hands, and elastic helps the fit of the glove.

Elasticated stitching is designed to mold to the grooves of the player's hand.

Breathable

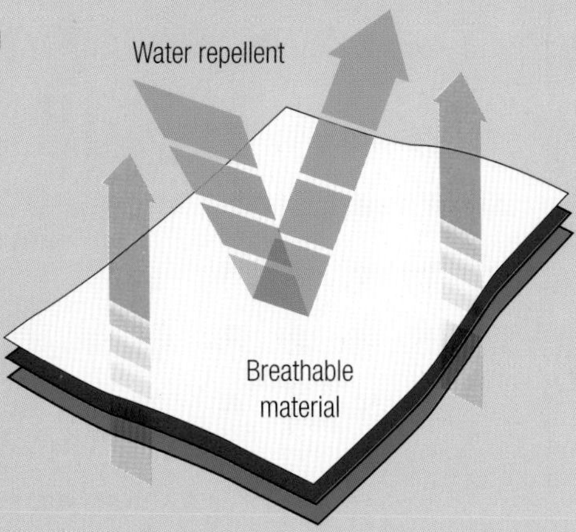

Water repellent

Breathable material

▲ **Gripping stuff** The modern golf glove is the accumulation of years of research-driven development. Many features, such strategically placed, ribbed contact points, meshed motion zones, and contoured stitching, create a glove that helps the golfer grip the club more comfortably.

▶ **Material improvement** Fabric technologies have revolutionized the performance of sports clothing, and the humble golf glove has benefited from such advancement. Today, glove materials are combined to provide a breathable, lightweight, and comfortable membrane that also repels external moisture.

What are the core movements that set Tour pros apart?

What movements help Tour pros to hit the ball better?

From watching the Tour pro on television, most people can easily recognize their swings as powerful and efficient. Not surprisingly, the bulk of biomechanical studies[1] have confirmed that professional golfers achieve faster body rotation and a better-timed sequence of the rotating segments—both of which lead to faster clubhead speed compared with high-handicappers. These differences between professionals and less-skilled golfers seem fairly obvious, but what exactly are the characteristics of the swing which lead to faster body rotation and, ultimately, faster clubhead speeds?

▼ *Tilt-shift*   *The image of the professional golfer (right) demonstrates two key characteristic core movement patterns at the top of the backswing. First, the thorax is tilted away slightly from the target (white line), keeping close to the address position, and second, the pelvis has maintained its lateral position during the backswing (blue line). These core movements are possible because of a balanced interaction with the ground. The professional golfer also shows reduced lateral shear forces and a constant pressure on the medial aspect of each foot (red lines). In contrast, the high-handicap golfer (left) has shifted the pelvis laterally (blue line), tilted the thorax toward the target (white line), and allowed foot pressure to move to the lateral aspect of the rear foot (red lines).*

Top of backswing

High-handicap golfer

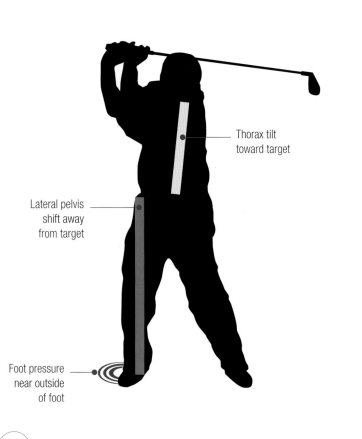

Thorax tilt toward target

Lateral pelvis shift away from target

Foot pressure near outside of foot

Professional golfer

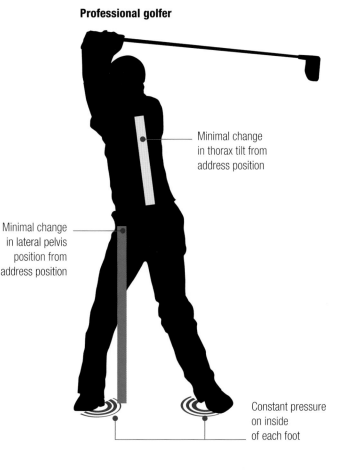

Minimal change in thorax tilt from address position

Minimal change in lateral pelvis position from address position

Constant pressure on inside of each foot

There are many finely tuned and highly coordinated movements of the arms, hands, and club that contribute to faster clubhead speed, but limb movements represent the end of the kinematic chain and these must be preceded by proper movements in the core—the pelvis and thorax. In terms of core movement, there are several key moves the professional golfers make that distinguish them from the average golfer. In particular, professional golfers maintain lateral stability and posture during the backswing to a greater extent. Core stability can be seen by looking for lateral tilting of the thorax and lateral shifting of the pelvis—professionals show minimal change in both of these key indicators during the backswing.[2,3] In the downswing, the professionals rotate their pelvis toward the target earlier

▼ **Read the hips** *A side-on view at impact: the professional golfer (right) has rotated the pelvis toward the target. The high-handicap golfer on the left has kept the pelvis square to the target at impact, limiting clubhead speed. In addition to rotating the pelvis toward the target, the professional has also shifted the pelvis toward the target during the downswing, evidenced by the right heel being slightly raised off the ground at impact.*

than the average golfer—this is commonly referred to as "clearing the hips." In fact, professional golfers begin the downswing with pelvis rotation, whereas the average golfer tends to begin the downswing with thorax and shoulder rotation. Finally, at impact, the professional's pelvis and thorax are both opened toward the target.[3]

These core movements are achieved as a consequence of the golfer's interaction with the ground. Professional golfers impart greater anterior–posterior shear forces and reduced lateral shear forces on the ground compared with less-skilled golfers. The skilled golfer's foot pressure is to the inside (medial) and toward the heel (posterior) of each foot.[4] This interaction with the ground promotes stability at the transition of the backswing. This then sets up a quick, powerful rotation during the downswing, which is initiated by the anterior–posterior shear forces. The average golfer struggling with low clubhead speed could learn a great deal from studying the core movements of professional golfers.

Impact

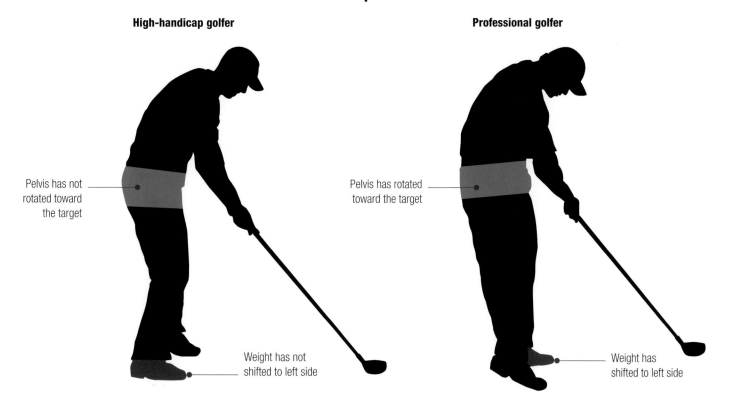

High-handicap golfer

Pelvis has not rotated toward the target

Weight has not shifted to left side

Professional golfer

Pelvis has rotated toward the target

Weight has shifted to left side

Is weight transference significant?

How much should I shift my weight during the swing?

Weight transference is a coaching term used to convey the proportion of the golfer's total weight that should be distributed over each foot during the swing. More accurately, biomechanists have defined weight shift as the change in the proportion of total downward force under the front foot throughout the movement.[1] According to popular coaching manuals, for full swings the weight should begin balanced evenly between both feet (about 50 percent each) at address.[1,2] Conventional coaching advocates that the weight should then shift toward the rear foot during the backswing and then rapidly toward the front foot in the downswing, and research seems to support this. However, other patterns of weight transfer have been identified which are associated with expert performance.

Two distinct weight-transfer techniques—the so-called "front-foot" and "reverse" styles—have been identified,[1] while a swing technique called "stack and tilt" is anecdotally popular among some players and coaches.[3] The front-foot style matches the conventional idea of weight transfer. The reverse style is distinguished by the weight shifting back toward the rear foot during mid-downswing. The stack-and-tilt swing, in terms of weight transfer, can be generalized as "staying centered over the ball" rather than shifting weight toward the rear foot during the backswing. Axial rotation and lateral tilt of the thorax achieve comparable angular speed profiles throughout the downswing.[4] Given that there is room for the thorax to tilt and the pelvis to shift laterally, this action would explain the weight shift toward the front foot in the swing. Furthermore, the peak in downward force occurring before impact (event ED) represents the force required for decelerating pelvis rotation and lateral movement, which is the beginning of the kinematic chain.[5] In light of these scientific findings, the reverse-style weight transfer appears to be associated with exceptional performance. An awareness of the two patterns presented here, and experimentation with pressure mats, will enable the player to explore which swing technique fits best.

Weight transfer styles

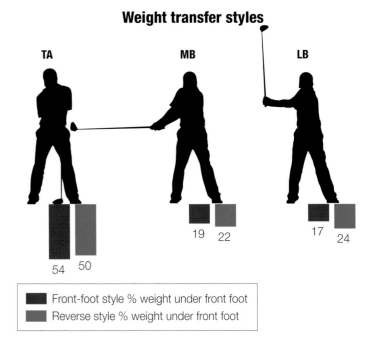

TA   MB   LB

19 22   17 24

54 50

■ Front-foot style % weight under front foot
■ Reverse style % weight under front foot

▼ *Feel the force Here the total force is visualized as a vector, with its length representing the magnitude of the force, and its direction showing the direction of the ground's reaction force to each foot. At TB, vertical force (weight) is greater under the rear foot (foot furthest away from target), shown by the longer rear foot vector. At ED, the rear foot vector points posteriorly, which means the rear foot is pushing anteriorly into the ground. The front foot (foot closest to the target) is pushing in the opposite direction (posteriorly). The opposing directions of the ground reaction forces by each foot cause rotation. At BC, the directions of both force vectors have reversed.*

Total force vectors

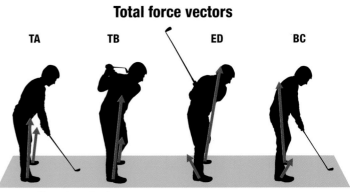

TA   TB   ED   BC

Ground reactive force

▼ Weight-watching *The nine events for a low-handicap golfer: take-away (TA), mid-backswing (MB), late backswing (LB), top of backswing (TB), early downswing (ED), mid-downswing (MD), ball contact (BC), follow-through (MF), finish position (FP).[1] The percentage bars* *represent average values for the proportion of total vertical force under the front foot (e.g. 0 percent = no weight under front foot, 100 percent = all weight under front foot). Red values correspond with the front-foot group and blue values with the reverse-style group.*

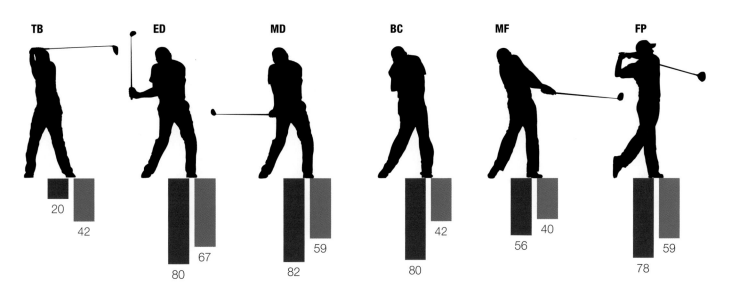

TB	ED	MD	BC	MF	FP
20	67	59	42	40	59
42	80	82	80	56	78

Front-foot style

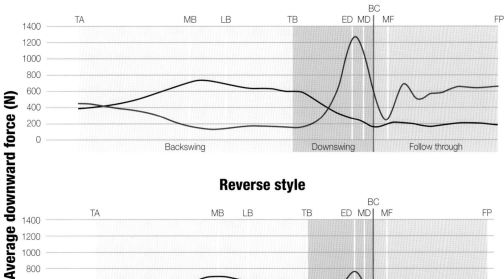

Average downward force (N)

TA · MB · LB · TB · ED MD · BC · MF · FP
Backswing · Downswing · Follow through

Reverse style

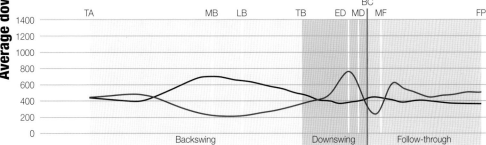

TA · MB · LB · TB · ED MD · BC · MF · FP
Backswing · Downswing · Follow-through

◄ Fancy footwork *The downward force for each of the golfer's feet is shown: red for the front foot, black for the rear foot. The top graph is an example of the downward forces involved with a swing consistent with the front-foot style, and the bottom one a swing consistent with the reverse style.[1] In the reverse group the weight starts to shift back toward the rear foot at ED. The peak in front-foot downward force (at ED) is linked to the start of the kinematic chain. Notice that the sum of vertical forces for both feet does not necessarily equal the golfer's weight (weight in N = mass in kg x acceleration due to gravity at 9.8 m/s^2). Because of angular momentum, the rotation can drive the golfer into the ground or lift the golfer off the ground. For this reason some players' heels leave the ground during ball contact.*

— Front foot downward force
— Rear foot downward force

assessing the risk

Being a good golfer is not just about being able to hit the ball well and consistently, but also about knowing which shots to make in which circumstances—making the right decision at the right time in the right place. This mostly means setting up each shot to give you the best chance of making par or better.

How a player decides on the most appropriate shot at any one time is governed by many factors. Is it best to play it safe, or take a risk? Do you "take on" the hazard—water or sand—or avoid it? How a player is performing against their handicap or an opponent, their position on the leader board, or their current mental state following a run of good or bad luck can all dictate whether a conservative or risk strategy is selected.

Although it seems obvious that a more able golfer would take fewer risks throughout their round, given their aptitude and the potential cost of the risks, and that a novice would have more to gain from a riskier strategy, this may only be partly true[1]: research does also reveal that our general risk-taking behavior throughout the 18 holes may say more about our character than our score.[2] Top golfers instinctively select a strategy for each shot based on the environmental cues they observe and their past experiences. Firstly, by scanning the hole they are consistently assessing the risks. They evaluate the probability of a particular shot type being successful, and assess the penalty if it goes wrong and the reward if they pull it off.

How we assess risk is very much dependent on what we have to lose if it all goes wrong. Even for players of a high standard, playing it safe does not mean avoiding the rough—or even hazards—as much as ensuring that the next shot will enable them to use a predictable swing, resulting in accurate length and line.

▶ *Risk assessment* *Assessing the risk of any shot comes down to two important factors: the likelihood of success and the severity of the penalty if it all goes wrong. Having assessed the risk, here Martin Kaymer of Germany plays a tee shot across a wide expanse of water during his first round of the French Open at Le Golf National.*

What is the optimal pattern of wrist torque for maximizing clubhead speed?

Can the "late hit" help my driving distance?

There is a lot of discussion about the release of the wrists during a swing as impact approaches. The so-called "late hit," popular in many coaching manuals, is characterized by keeping an acute angle between the lead forearm and the club shaft for as long as possible during the downswing. Cochran and Stobbs proposed that although the "natural wrist release"—caused by centrifugal force of the swinging club—provides plenty of clubhead speed, slightly more could be squeezed out of the swing if the wrists actively applied torque late in the downswing.[1] They also noted that in order for this strategy to work the wrist angle would have to be released later than in the natural release.

More recently, many studies have looked into optimizing the release behavior of the wrists before impact. In general, three main wrist release strategies have been investigated: the "natural release," which requires no muscular torque; a "delayed wrist action," in which wrist torque is applied to maintain the wrist angle and promote the late hit; and the "delayed-active wrist action," similar to the delayed wrist action but followed by an active wrist torque to accelerate the club into the ball. Simulation studies have agreed that the delayed-active torque technique offers potential for maximizing clubhead speed.[2,3,4] Sprigings and Mackenzie showed that if wrist torque is applied to maintain the wrist–shaft angle until the lead arm reaches "eight o'clock" when viewed from the front (the "delayed wrist action"), at which point the torque is applied in the opposite direction to actively release the wrists, clubhead speed could be maximized (by about 1.6 percent faster than natural release). However, the timing window in which these torques need to be applied is extremely sensitive, which suggests that the benefits associated with delayed-active wrist torque are difficult to achieve. If the active torque is mistimed by just 50 ms (half the time it takes to blink), clubhead speed would be slower than the natural release. This sensitivity to timing suggests optimizing wrist action is highly dependent on the player's proprioception, or feel. All of which suggests that, for the average golfer, it is probably more worthwhile keeping a relaxed grip and trying to release naturally.

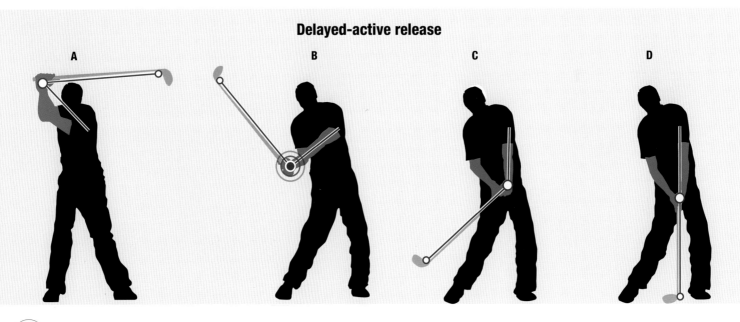

Delayed-active release

A   B   C   D

Delayed wrist action

▼ **Delayed and natural releases** *The sequence on the left represents a swing with the delayed-active (optimal) wrist torque from Sprigings and Neal.[4] The wrists release because of the centrifugal force acting on the club, and at "B" muscular torque is applied to further release the wrists and accelerate the club. On the right, no muscular torque is applied to release the wrists, only the centrifugal force acting on the club. In simulation the delayed-active release reached a clubhead speed of 98.4 mph (158.4 km/h) and the natural release 90.4 mph (145.5 km/h).*

▲ **Optimized release** *The swing sequence shows the optimal wrist release in the drive. When the left arm is at "nine o'clock" muscular force is applied to maintain the wrist angle. Once the left arm reaches about "eight o'clock" the wrists should forcefully release the club through ball contact. Both wrist torques must work in concert with each other—if the wrist angle is actively released too early the advantage of applying wrist torque is lost.*

Natural release

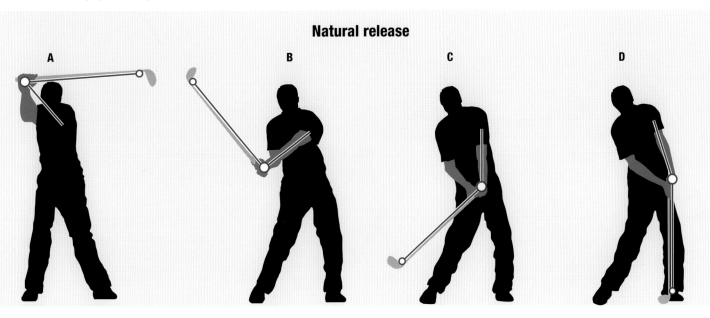

A   B   C   D

How do grip forces change during the golf swing?

How firm should my grip be?

The grip is one of the most important facets of the golf swing because it represents the only interface between the golfer and the club through which all force and energy must be transmitted. Many golf instruction manuals recommend that the grip should be kept as loose as possible to allow a full and unrestricted release of the golf club, but also firm enough to prevent the club from slipping, particularly during the impact phase. A loose grip can also help to maximize the "feel" of a golf shot and reduce the likelihood of overuse and repetitive strain injuries occurring in the hands, wrists, and forearms.

Another recommendation that appears to be prevalent in the golf coaching literature is that grip firmness should remain fairly constant throughout the golf swing.[1] However, a basic mechanical analysis indicates that this supposition cannot be correct. When a golf club is swung, centrifugal forces are generated, which are proportional to the amount of acceleration and deceleration of the golf club. These load forces have been estimated to be up to 450 N for very fast swings.[2] Additional load forces are also generated during the impact phase through collision between the clubhead and the golf ball and turf, the magnitudes of which depend on their respective physical characteristics and relative momentum at impact. Grip forces, therefore, need to be adjusted simultaneously or slightly ahead of fluctuations in load forces to avoid an inefficient transfer of force and energy as a result of the golf club slipping.

Owing to difficulties with measuring grip forces, only a few biomechanical investigations have focused on this aspect of the golf swing. Budney[3] measured grip forces under the last three fingers of the left hand, on the base of the first three fingers of the right hand, and under the left thumb, using an instrumented driver fitted with strain gauges. Although individual differences were reported, the three golf professionals analyzed exhibited grip forces which were closer to those recommended in the coaching literature than did the three amateur golfers. However, owing to the small number of golfers analyzed in this study, the generality of these findings is limited. In a more extensive recent study, Komi and colleagues[4] showed that 20 golfers (male and female), regardless of playing ability, had their own unique grip force "signature"—that is, grip forces were highly consistent for each golfer over repeated shots but varied considerably between golfers. Certain trends were evident across golfers, however, such as local peaks in grip force just before and after impact, and an overall higher force on the left hand than the right hand for nearly all golfers tested. The data here are for right-handed golfers but it is worth noting that the situation is mirrored for left-handed golfers.

Measuring grip forces

◀ *Handy hints* Force sensors use a semi-conductive ink that is applied between electrical contacts and thin polyester sheets, giving the sensors a resultant thickness of just 0.1 mm. These extremely thin and lightweight sensors can be fitted around the circumference of the golf grip, or attached to the gloves themselves, to allow the changing forces at numerous locations to be measured simultaneously.

Grip forces

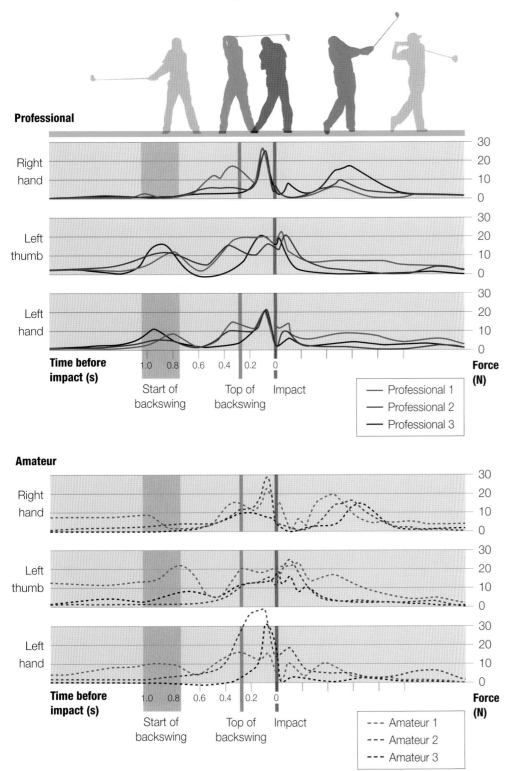

Professional

Right hand

Left thumb

Left hand

Time before impact (s)

1.0 0.8 0.6 0.4 0.2 0

Start of backswing

Top of backswing

Impact

Force (N)

30 20 10 0

—— Professional 1
—— Professional 2
—— Professional 3

Amateur

Right hand

Left thumb

Left hand

Time before impact (s)

1.0 0.8 0.6 0.4 0.2 0

Start of backswing

Top of backswing

Impact

Force (N)

30 20 10 0

--- Amateur 1
--- Amateur 2
--- Amateur 3

◀ *Getting a grip* Even as far back as the late 1970s, scientists were able to measure grip forces throughout the swing by using a modified grip equipped with a series of strain gauges. Results collected from professional and amateur players began to reveal demonstrable differences in force patterns between skill levels.[3]

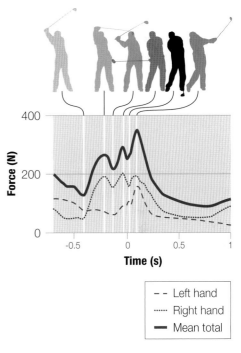

Force (N)

400

200

0

-0.5 0 0.5 1

Time (s)

--- Left hand
···· Right hand
—— Mean total

▲ *In the swing* Analyses of grip forces using grip and glove sensors in conjunction with high-speed video have revealed highly individualized grip force patterns across golfers regardless of playing standard.[4] An example force–time profile for the left and right hand, and both hands combined, is shown here, along with six key points and corresponding swing position images.

59

Can biomechanical analyses help to increase consistency?

Can the "swing plane" improve how I hit the ball?

The "swing plane" concept was originally introduced by Seymour Dunn in the 1920s but popularized by Ben Hogan during the 1950s in his classic instructional text, *Five Lessons: The Modern Fundamentals of Golf*.[1] The swing plane is now generally considered to be an imaginary two-dimensional surface, extending from the center of the golf ball through the top of the golfer's sternum, along which the clubhead should travel during the swing. It is thought to be an important aid to improving the swing because it supposedly enables the golfer to deliver the club more consistently at impact, leading to less dispersion of shot outcomes.

There have been several biomechanical investigations into the swing plane concept but these have typically been equivocal and contradictory. For example, in three-dimensional kinematic studies by Vaughan and Neal and Wilson, it was found that the plane of the shaft of the golf club was not constant for any substantial period of time during the golf swing. In contrast, Lowe and Fairweather reported that the downswing and follow-through phases of the swings of expert golfers were approximately planar. More recently, Coleman and Anderson[5] showed that it was possible to fit a single plane to the motion of the golf club during the downswing for a group of experienced golfers, but the fit varied between golfers and also between clubs (that is, a pitching wedge, a 5-iron, and a driver).

Despite being the focus of a number of scientific investigations, the swing plane still remains a contentious concept and further empirical research is warranted. However, Jenkins[1] has suggested that the concept might be used as an "idealized replica" to help golfers develop a mental image of what their golf swing should look like in their quest to improve. Indeed, in his original writings, Hogan noted great improvements in the techniques and performances of golfers that he had personally taught after encouraging them to visualize the backswing and downswing movements along the swing plane.

◀ **Plane simple** *Ben Hogan famously visualized the swing plane as a large pane of glass running from the ball parallel to the target line and placed over the golfer's head so that it rests on their shoulders, the theory being that the shoulders, arms, and hands should all move along this plane throughout much of the golf swing. The angle of inclination of the swing plane is determined by the golfer's stature and the distance they stand from the ball at address, which is dictated by the club used (for instance, a pitching wedge, 5-iron, or a driver; see above).*

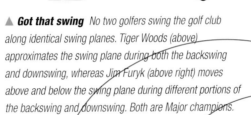

▲ **Got that swing** *No two golfers swing the golf club along identical swing planes. Tiger Woods (above) approximates the swing plane during both the backswing and downswing, whereas Jim Furyk (above right) moves above and below the swing plane during different portions of the backswing and downswing. Both are Major champions.*

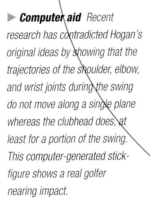

▶ **Computer aid** *Recent research has contradicted Hogan's original ideas by showing that the trajectories of the shoulder, elbow, and wrist joints during the swing do not move along a single plane whereas the clubhead does, at least for a portion of the swing. This computer-generated stick-figure shows a real golfer nearing impact.*

Is there evidence for an optimum movement model?

→ Can a perfect swing be achieved?

It is tempting to treat the golf swing technique as an optimization problem, in that every position the golfer moves through has an ideal. This is the approach Mann and Griffin[1] attempted to support by combining the swings of over 100 professional golfers. However, while it might be theoretically possible to create a simulation or robot capable of such a perfect, repeatable sequence, the human player has many more variables to contend with, both internal and external, so that a single, fixed swing pattern is not what's required. In other words, each player must consider their own strengths and weakness, which have arisen from their past experience and can be psychological as well as physiological. The player must then use that knowledge to simplify the task of coordinating nearly 800 separate muscles, which must be precisely controlled to create a functional swing.[2]

Storing the vast number of configurations into which our body segments could be arranged theoretically requires computational power beyond the brain's neurophysiological limits—and yet, of course, humans are able to coordinate highly complex patterns of movement, including the golf swing, with apparent ease and fluidity. A leading explanation for our ability to overcome the seemingly infinite number of possible swings to produce a functional swing comes by way of synergies.[3]

A synergy is a conceptual linkage of parts (such as muscles) that reduces the information the brain needs to supply to operate the movement. When coaches refer to "moves" in the golf swing (for example, lead the downswing with the hips) they are supplying a small amount of information that collectively represents a great deal of information. Synergies operate in a similar way. What makes a golf swing good is not just the positions it moves through but, probably more importantly, the stability of these temporary synergies which are very individual-specific and, if stable, will lead to predictable golf shots.

Biomechanical studies confirm that individual variation is the rule rather than the exception and, although swings of good golfers may have certain commonalities, each golfer still has their own signature.[4,5] It may be reasonable to imagine a perfect swing for an individual playing a specific shot, but throughout a round of golf, a tournament, or a career, to perform well the individual must be able to adapt the swing to suit the specific situation and the shot being played. Instead of a perfect swing, it is probably better to talk about an adaptable swing that adheres to general biomechanical principles and can accommodate many different situations. Practice drills which improve kinesthetic awareness are probably very influential in improving golf-specific synergies.[6]

Creating a functional swing

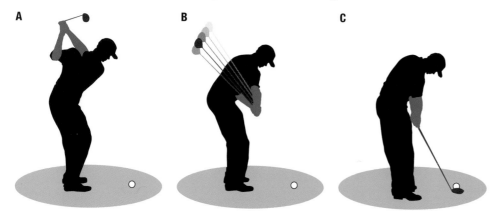

A   B   C

◀ **How a golf swing works** *One approach to explaining a functional swing is shown here:[7] the golfer uses a previous state (A) to predict a future state in the swing (B); when the golfer arrives at the predicted state (B), sensory information is compared with the expected information and used to make the necessary adjustments for future states (i.e. body position and velocity at impact —C). This model allows for uncertainty at the intermediate stage in the movement (B), after which error is reduced at the critical instant (time of impact) given a known target (ball location).*

Model movement

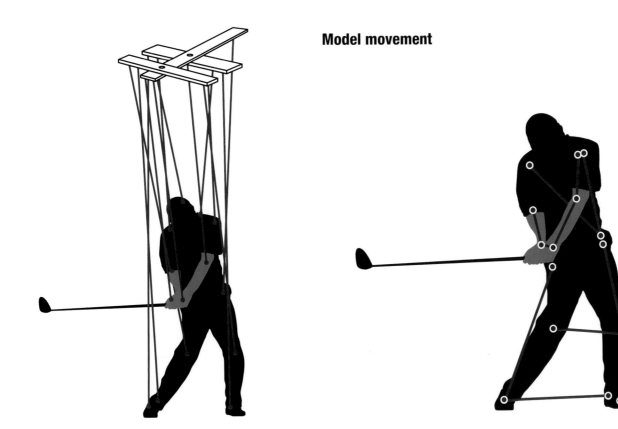

▲ **Concepts of swing control** The left-hand image illustrates the concept consistent with the "perfect swing" school of thought: the individual body parts must be controlled independently by the brain. The right-hand image shows the alternative concept: synergies are created and controlled by the brain as a unit rather than individual parts. If one part is "out of position" its linkage to other parts makes compensatory movements possible. Notice also how the interactions in the second image reduce the amount of information needed by the brain to control the golf swing. These images are adapted from the original conception by Michael Turvey.[2]

▼ **Measuring synergies** Two variables recorded at impact in the golf swing can be plotted against each other to show whether they represent a synergy. Elongated clusters of data points along the diagonal line indicate a compensatory movement—a synergy: if one component of the swing is out of position (if its value is too high or too low) the other variable will compensate to achieve roughly the same outcome. In some cases, a certain amount of variability in the golf swing is a good thing because it affords the golfer flexibility and adaptability for new and unexpected situations.

Compensatory movements

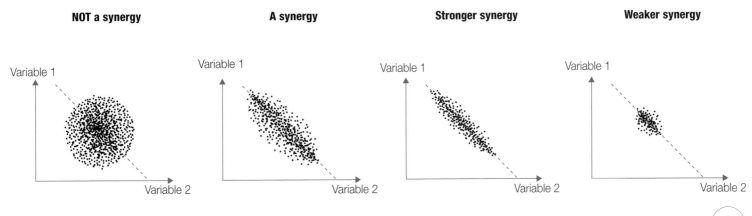

While a golf club is built from only three components, the technical elements of its design and performance are as varied as the golfers who play the game. Golfers vary tremendously in their size, strength, athletic ability, and swing characteristics. A golf club with particular technical specifications will perform completely differently in the hands of different golfers. Matching each golfer with the right golf equipment that will enable them to play to the best of their ability is a specialized field within the game, called clubfitting. Accurate clubfitting requires a scientific understanding of how the different swing characteristics of golfers interact with the technical differences in golf equipment to result in differences in shot performance. In this chapter, Richard Kempton reveals why technical performance differences occur in golf equipment and how they are selected for each golfer, bringing the science of equipment right into the 21st century.

chapter three

the equipment

Richard Kempton

Is there an optimum driver length?

→ Will I hit the ball farther with a longer driver?

Probably one of the biggest myths in the game is that a longer driver will *automatically* produce more distance than a shorter one for all golfers. A longer driver certainly can result in longer drives, but it's definitely not a universal truism. Ultimately the distance any golfer is theoretically capable of driving a golf ball will depend on how hard they hit it (i.e. clubhead speed at impact). Whether or not that will actually result in maximum distance depends on whether the impact results in the golf ball leaving the clubface at the highest possible velocity, combined with the optimum launch angle, and with the optimum amount of spin to produce the optimum trajectory and landing angle for the ground conditions. There is no single optimum combination of launch angle and spin that will produce maximum distance for all golfers, because that will vary with initial ball speed and angle of attack (the upward or downward path of the clubhead at impact).

A longer shaft should produce more clubhead speed by virtue of the increased radius of the arc described by the clubhead. It's even possible to calculate how much extra clubhead speed an extra 1–2 in (2.5–5.0 cm) of shaft length will theoretically produce. However, golfers react to the weight and balance (assembled moment of inertia or swingweight) of a club, the way it feels, and even the way it looks, so—in practice—a longer shaft may not result in more clubhead speed. A longer shaft can even reduce clubhead speed rather than increasing it, if its swingweight (or more correctly the moment of inertia of the assembled club about the end of the shaft, which coincides roughly with the wrist joint of the upper hand) is greater than the golfer can swing and release efficiently. As a result, when driver lengths are increased, it is normally necessary to reduce the clubweight—usually most easily achieved by installing a lighter shaft. Assuming that the driver head is of the correct loft and configuration to provide the optimum launch angle, spin rate and trajectory, increasing driver distance is actually about producing more ball speed, and that doesn't necessarily correlate with clubhead speed.

Factors affecting drive distance

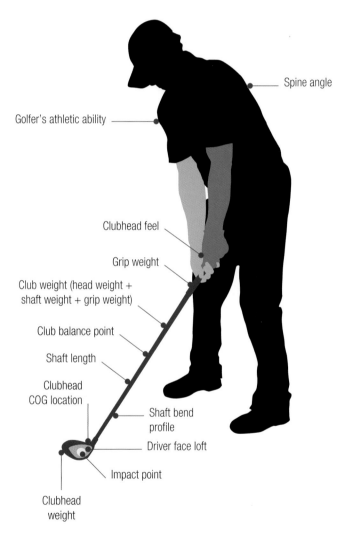

Spine angle

Golfer's athletic ability

Clubhead feel

Grip weight

Club weight (head weight + shaft weight + grip weight)

Club balance point

Shaft length

Clubhead COG location

Shaft bend profile

Driver face loft

Impact point

Clubhead weight

▲ *Finding balance* *There's more to driving distance than driver length. Its "balance" (swingweight or moment of inertia) is affected by a number of other factors: clubhead weight, the clubhead center of gravity relative to the end of the grip, the weights and balance points of the shaft and grip, and even the ratio of the clubhead weight to the shaft weight.*

Smash factor and distance

CLUB SPEED (mph)	SMASH FACTOR (initial ball speed divided by clubhead speed)															
	1.50	1.49	1.48	1.47	1.46	1.45	1.44	1.43	1.42	1.41	1.40	1.39	1.38	1.37	1.36	1.35
	INITIAL BALL SPEED (mph)															
90	135.0	134.1	133.2	132.3	131.4	130.5	129.6	128.7	127.8	126.9	126.0	125.1	124.2	123.3	122.4	121.5
91	136.5	135.6	134.7	133.8	132.9	132.0	131.0	131.1	129.2	128.3	127.4	126.5	125.6	124.7	123.8	122.9
92	138.0	137.1	136.2	135.2	134.3	133.4	132.5	131.6	130.6	129.7	128.8	127.9	127.0	126.0	125.1	124.2
93	139.5	138.6	137.6	136.7	135.8	134.9	133.9	133.0	132.1	131.1	130.2	129.3	128.3	127.4	126.5	125.6
94	141.0	140.1	139.1	138.2	137.2	136.3	135.4	134.4	133.5	132.5	131.6	130.7	129.7	128.8	127.8	126.9
95	142.5	141.6	140.6	139.7	138.7	137.8	136.8	138.9	134.9	134.0	133.0	132.1	131.1	130.2	129.2	128.3
96	144.0	143.0	142.1	141.1	140.2	139.2	138.2	137.3	136.3	135.4	134.4	133.4	132.5	131.5	130.6	129.6
97	145.5	144.5	143.6	142.6	141.6	140.7	139.7	138.7	137.7	136.8	135.8	134.8	133.9	132.9	131.9	131.0
98	147.0	146.0	145.0	144.1	143.1	142.1	141.1	140.1	139.2	138.2	137.2	136.2	135.2	134.3	133.3	132.3
99	148.5	147.5	146.5	145.5	144.5	143.6	142.6	141.6	140.6	139.6	138.6	137.6	136.6	135.6	134.6	133.7
100	150.0	149.0	148.0	147.0	146.0	145.0	144.0	143.0	142.0	141.0	140.0	139.0	138.0	137.0	136.0	135.0

▲ *Effect on distance* A longer driver can increase distance, but only if it enables you to increase your clubhead speed *and* your ball speed. That will depend on what happens to your smash factor when you swing a longer driver. If your smash factor improves—or the fall in smash factor is minimal—and you are able to generate more clubhead speed (and you can maintain or improve your angle of attack, launch angle, and spin rate), you will see a distance increase (but possibly not as much as you think). However, if your smash factor drops too much (or your angle of attack, launch angle, and spin rate change for the worse), you could actually lose distance. This table helps to show why: to convert an increase or decrease in *ball* speed to an approximate yardage change, multiply by a factor of 1.7–1.8. (Note that for many golfers, a 1 in (2.5 cm) longer driver will not increase their clubhead speed by more than 2–3 mph/3–5 km/h, unless the total club weight is also significantly reduced.)

A longer driver, even one fitted with a lighter shaft, may—and usually will—result in the ball being struck less consistently on face center and less squarely, which will affect the ball speed to clubhead speed ratio (the "smash factor"). For every golfer there will come a point at which the law of diminishing returns starts to operate, where (a) a longer driver gives more clubhead speed, but the smash factor drops to a level which results in no increase in ball speed and distance, or even a reduction in ball speed and distance; or (b) a longer driver results in less clubhead speed and less ball speed and distance. That compromise is ultimately a matter of golfer choice, but remember that it is easier to score from the fairway than the rough.

Ball speed = (clubhead speed) x (smash factor)

In the club

Driver fitting is a complex balancing act between club length, shaft and club weight, club balance, clubhead loft, and clubhead face angle and feel. All these elements affect clubhead speed and the ability to hit the ball consistently on face center, as well as the launch angle, spin rate, trajectory, and spin axis tilt.

▲ *Impact zone* As the ball impact shifts away from the center of the clubface, both the coefficient of restitution and the smash factor will decrease, resulting in reduced distance and accuracy.

How far can a golfer drive the ball (step 1)?

How can I maximize my drive distance?

For any combination of altitude, ground and weather conditions, the maximum distance any golfer should theoretically be able to hit a driver can be estimated reasonably accurately from their clubhead speed and a knowledge of ballistics. On normal, level fairways in completely still conditions, at sea level, there is no technology permissible under the Rules of Golf that will enable a golfer with a sub-90 mph (145 km/h) clubhead speed to drive the ball 300 yd (275 m). However, in order for a golfer to achieve maximum distance, all three of the following conditions will need to be met: maximum clubhead speed is attained; the ball is struck on the area of the face where the coefficient of restitution is highest to maximize the smash factor and thus the initial ball speed; the combination of the launch angle (the angle at which the ball leaves the clubface relative to the horizontal) and its initial spin rate provide the best possible trajectory and landing angle for that initial ball speed. As a result, maximizing driver distance requires a two-pronged approach.

Step 1: optimizing for clubhead speed and impact consistency

The length, total weight, and "club balance" of a driver (its swingweight or moment of inertia) all potentially affect clubhead speed, but in different ways—and as stated above, simply maximizing clubhead speed will not automatically produce maximum distance unless both the initial ball speed and the trajectory are also optimal. The same three things also affect a golfer's ability to impact the ball squarely and on face center and the degree of accuracy and consistency they will achieve with it, so there is almost always a need to strike a compromise between maximum ball speed and acceptable accuracy. That compromise is ultimately a matter of golfer choice, but remember that it is easier to score from the fairway than the rough.

A lighter driver can allow a golfer to increase their clubhead speed and thus potentially hit the ball farther but, for many golfers, the distance increase will not be significant unless the club total weight is reduced by 20 g or more compared with their current driver—but accuracy/consistency may suffer, unless the assembled swingweight or moment of inertia is maintained at the level that the player needs.

In very general terms, increasing the length of a driver by an inch could potentially add 7–8 yd (6.4–7.3 m) to a golfer's driving distance—provided that they can still maintain their smash factor at the same level as with a shorter club and its total weight, swingweight or assembled moment of inertia do not increase above the thresholds that the golfer can consistently handle, and accuracy and consistency are still acceptable.

How do you know which combination of club length, total weight, and swingweight/moment of inertia will do all these things? Your best option is to find a competent clubfitter who can not only work all that out for you, but also take care of the second aspect of maximizing your driving distance: determining the optimum combination of driver head and loft, shaft flex, and shaft bend profile for the way you swing the club (i.e. your clubhead speed and angle of attack) and the conditions on the courses you generally play.

The two tables opposite show clearly that, for a given clubhead speed (assuming that the club is set up to allow the golfer to achieve the highest smash factor possible with a conforming driver), the combination of launch angle and spin rate needed for maximum distance depends not only on angle of attack (AOA), but also on whether the driver is to be optimized for maximum *carry* distance or maximum *total* distance.

Maximizing distance

*The tables (right) show the distances that should be achievable for range of clubhead speeds (column 1) and angles of attack (column 2), depending on whether it is fully optimized for maximum **carry** (top table) or maximum **total** distance (bottom table). With a "Tour" type ball, in still conditions at sea level and on normal fairways, the launch angles (column 3) and spin rates (column 4) will produce the distances given in columns 5 and 6 when the dynamic loft at impact is as shown in column 7. Column 7 "translates" the dynamic lofts into actual driver lofts (to the nearest 0.5 degrees), provided that all three of the following assumptions are true: 1) on-center impact, (2) shaft bending adds 1.5 degrees of loft dynamically at impact, and (3) the golfer's swing mechanics add NO loft dynamically. (Do not assume that the actual loft of any driver will be as marked unless it has been checked in a gauge.) If a golfer with a 90 mph (145 km/h) clubhead speed opts to have a driver optimized for maximum **total** distance (bottom table), the data shows that he will need about 5.5 degrees more driver loft with a −5 degree (negative or downward) angle of attack than if his attack angle were +5 degrees (positive or upward). The combination of a more upward attack angle and a lower driver loft will produce just over 25 yd (22.9 m) more carry and almost 30 yd (27 m) more total distance than the higher loft/negative AOA combination.*

Both tables are adapted from data collected and analyzed by the makers of the TrackMan™ launch monitor, Trackman a/s, Denmark (www.trackman.dk).

Optimization for maximum carry distance[1]

Clubhead speed (mph)	Attack angle (degree)	Launch angle (degree)	Spin rate (rpm)	Carry distance (yd)	Total distance (yd)	Dynamic loft (degree)	Approx driver loft (yd)
75	−5	14.6	3722	143	166	18.2	21.5
	Level	16.3	3121	154	178	19.2	17.5
	+5	19.2	2720	164	187	21.8	15.5
80	−5	12.9	3652	160	176	16.2	19.5
	Level	15.5	3179	171	187	18.3	17.0
	+5	18.0	2648	181	197	20.3	14.0
85	−5	11.9	3669	175	199	15.0	18.5
	Level	14.5	3164	187	211	17.1	15.5
	+5	17.0	2596	197	223	19.1	12.5
90	−5	11.1	3689	191	215	14.0	17.5
	Level	13.4	3093	203	228	15.8	14.5
	+5	16.4	2633	214	239	18.5	12.0
95	−5	9.9	3626	207	233	12.6	16.0
	Level	12.7	3114	219	244	15.0	13.5
	+5	15.7	2595	231	256	17.6	11.1
100	−5	9.6	3722	222	244	12.2	15.5
	Level	12.1	3118	235	258	14.3	13.0
	+5	14.9	2538	247	272	16.7	10.0
105	−5	8.7	3645	237	260	11.1	14.5
	Level	11.2	3038	251	275	13.2	11.5
	+5	14.5	2563	263	288	16.2	9.5

Optimization for maximum total distance[1]

Clubhead speed (mph)	Attack angle (degree)	Launch angle (degree)	Spin rate (rpm)	Carry distance (yd)	Total distance (yd)	Dynamic loft (degree)	Approx driver loft (degree)
75	−5	11.8	3214	140	182	14.9	18.5
	Level	13.0	2506	147	195	15.3	14.0
	+5	15.3	1976	156	206	17.1	10.5
80	−5	10.1	3078	154	188	12.8	16.5
	Level	12.1	2494	163	199	14.3	13.0
	+5	14.8	2005	174	209	16.5	10.0
85	−5	9.3	3110	169	215	11.9	15.5
	Level	11.7	2568	180	228	13.8	12.5
	+5	14.0	1964	189	241	15.6	9.0
90	−5	8.5	3122	185	231	11.0	14.5
	Level	10.8	2517	196	245	12.8	11.5
	+5	13.8	2021	207	259	15.3	9.0
95	−5	7.9	3144	201	247	10.2	13.5
	Level	10.5	2565	213	262	12.3	11.0
	+5	13.0	1948	223	276	14.4	8.0
100	−5	7.2	3118	216	262	9.3	13.0
	Level	10.0	2570	230	278	11.7	10.0
	+5	12.4	1887	239	293	13.7	7.0
105	−5	6.4	3071	231	278	8.4	12.0
	Level	9.1	2461	243	294	10.7	9.0
	+5	11.7	1810	254	309	12.9	6.5

How far can a golfer drive the ball (step 2)?

How can I maximize my drive distance?

Step 2: optimizing for launch angle and spin rate

In general terms:

1 The lower your clubhead speed, the more loft you will need for maximum distance, and vice versa.

2 For a given clubhead and ball speed, as launch angle increases, the spin rate will need to decrease for maximum distance, and vice versa.

3 Launch angle is controlled by the loft of the clubface at the point of impact (its "dynamic" or "effective" loft), which includes the golfer's angle of attack. Spin is controlled by a combination of "spin loft" (loft at impact relative to the angle of attack), clubhead speed, and vertical gear effect (see page 89).

4 Dynamic loft (also known as "effective loft") is not the same as the stated loft of the club. The latter is the actual loft at address or in a gauge (its "static loft").

5 Angle of attack changes affect launch angle but have little or no effect on the spin rate. Dynamic loft changes affect both launch angle and spin rate.

6 The higher the clubhead speed and the closer to impact the golfer releases the wristcock, the more the shaft will tend to be bowed forward (adding loft dynamically, hence the name), and vice versa. Estimates vary widely as to the maximum amount of loft that shaft bending can add, but 2.5–3.0 degrees seems to be generally accepted.

7 For a golfer with a late release, a shaft that is stiffer overall and/or has a more tip-stiff bend profile can produce significantly lower launch angles and spin rates, and vice versa.

8 A golfer with an early release will see less or even no noticeable trajectory change from different shaft flexes or bend profiles.

9 Dynamic loft can be—and often is—added by a golfer's swing mechanics, usually as a result of "flipping" the hands and the club through impact (see graphic on page 73); the reverse can also happen, but only in extremely rare cases.

▼ *Increasing dynamic loft To maximize the distance of a drive, the golfer has to achieve the ideal combination of ball speed, launch angle, and spin rate (below). An increase in dynamic loft (bottom right) will result in an increase in launch angle, but because the "spin loft" (the loft at impact relative to the AOA) has not increased, neither will the spin rate. If the "spin loft" increases, that will raise the launch angle, spin rate, and trajectory, which will affect the distance that the ball carries and rolls. These images are adapted from an original conception by TrackMan™.[1]*

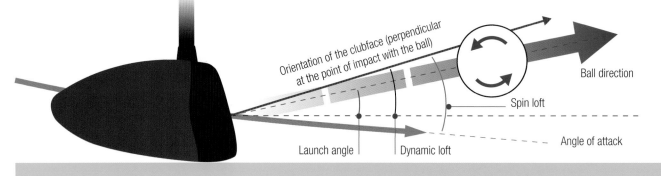

Orientation of the clubface (perpendicular at the point of impact with the ball)

Ball direction

Spin loft

Angle of attack

Launch angle

Dynamic loft

10 With drivers only, impacting the ball above the head center of gravity will reduce the spin rate and vice versa.

Step 1 should have determined the combination of club length, shaft, club total weight, and "swinging balance" (swingweight or moment of inertia) that provided the best combination of clubhead speed/ball speed, impact consistency—and feel. Some golfers are very sensitive to feeling how a shaft bends during the swing and have very distinct preferences for one shaft over another. Since they also tend to be the sort of player for whom different shafts can produce different trajectories and even precipitate unwanted swing changes, establishing the head specifications that will optimize their trajectory is probably best left until last.

If you have gained sufficient experience, as either a clubfitter or player, you will probably know instinctively if the ball flight you see is "right," but if not the data produced by an accurate launch monitor will certainly help to ensure that you get as close as possible to maximizing your potential distance with an acceptable level of accuracy and consistency. Even if a driver head is adjustable, typical manufacturing tolerances mean that there cannot be only very distinct differences between ostensibly identical heads, but also that the various settings will not necessarily correspond exactly to the published specifications.

▼ ***Better balance*** *These two illustrations demonstrate the importance of a properly balanced and fitted club. The left-hand image shows the impact pattern (after **six** shots) with a driver that is too long and the wrong balance for the golfer to control and swing well and consistently. The right-hand image shows the impacts (after **ten** shots) for the same golfer with the same driver, but this time customized for the best length and balance.*

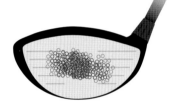

Incorrectly balanced club—impact pattern after **six** shots

Correctly balanced club—impact pattern after **ten** shots

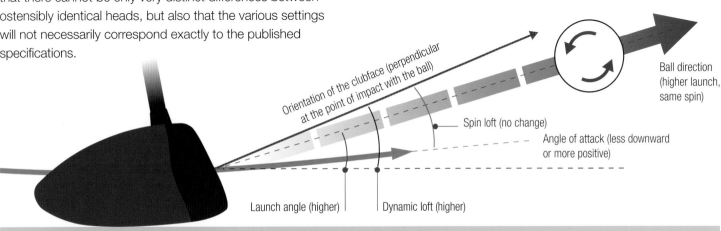

Orientation of the clubface (perpendicular at the point of impact with the ball)

Ball direction (higher launch, same spin)

Spin loft (no change)

Angle of attack (less downward or more positive)

Launch angle (higher)

Dynamic loft (higher)

What does the "kick point" reveal about how a shaft will perform?

How does the bend of a club affect my shots?

"Kick point" and "bend point" are thought to be reliable indicators of how high (or low) a trajectory the club shaft will produce. They tend to be used interchangeably, although they are measured in different ways: one involves compressing the two ends of the shaft between two steel plates, and the other clamping the butt end of the shaft and deflecting the tip end with a weight. The point of maximum bending is then measured relative to the tip in each case. Neither of these methods of measurement truly reflects how a shaft bends in response to the forces applied during the golf swing—and there are no agreed standards as to what constitutes a high, mid, or low kick point or bend point shaft (just as there are no agreed standards for the various flex designations used by different companies—for instance, L, A, R, S, Regular, Firm, or Lite).

A much more reliable indicator of the relative trajectories two or more shafts will produce is to compare what is called their "bend profiles" (the way in which the flex of each shaft is distributed along its length), by measuring each shaft's actual stiffness in either frequency (cycles per minute) or engineering EI units at multiple points along its length. As a side benefit, a shaft's bend profile will also suggest what swing type it will be likely to suit, in terms of clubhead speed, backswing to downswing transition (aggressive or smooth), swing tempo (fast or slow) and the release of the wristcock angle (early or late). However, golfers do not generally have access to bend profile data, which are difficult to interpret without quite lot of knowledge, so they still tend to think in terms of kick point or bend point, which are not necessarily reliable indicators of what sort of trajectory a particular shaft will provide, for two reasons: as already indicated, both terms are subjective; and whether or not two shafts with different flex distributions will actually produce visibly different trajectories depends to a large extent on how late or early the golfer releases the club (that is, where the wristcock angle starts to unwind).

Typical carbon fiber shaft

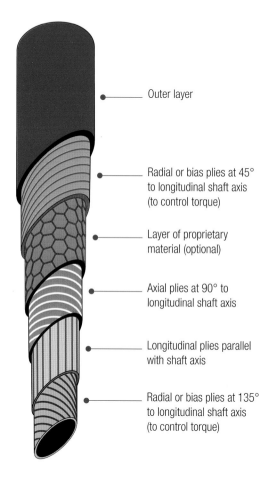

- Outer layer
- Radial or bias plies at 45° to longitudinal shaft axis (to control torque)
- Layer of proprietary material (optional)
- Axial plies at 90° to longitudinal shaft axis
- Longitudinal plies parallel with shaft axis
- Radial or bias plies at 135° to longitudinal shaft axis (to control torque)

▲ **Controlling torque** *The early graphite shafts had very low torsional stiffness (high torque), which was subsequently solved by incorporating several layers of "radial" or "bias" plies set at 45 degrees/135 degrees to the shaft axis. By varying the orientation, modulus (strength), and placement of the carbon fibers in the various layers that make up a shaft, its bending and torsional properties can be varied at different points.*

Shaft bend profiles and ball flight

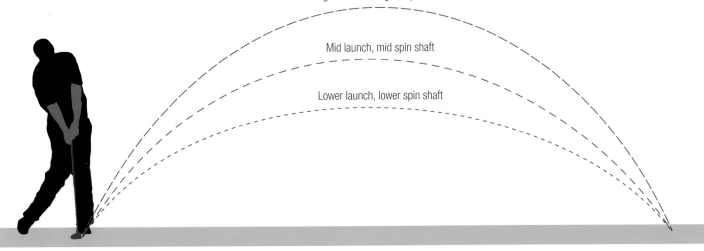

Higher launch, higher spin shaft

Mid launch, mid spin shaft

Lower launch, lower spin shaft

▲ **Trajectories from different bend profiles**
How much trajectory change any golfer will see from shafts with different bend profiles will depend on the clubhead speed, whether they release their wristcock angle early or late, and how much the bend profiles differ.

Release

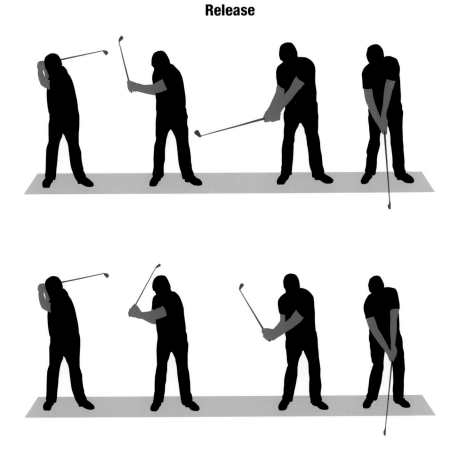

◄ **Effects of different releases** The golfer in the top sequence has an "early" release. He will tend to hit the ball high because he has allowed the clubhead to get ahead of his hands at impact by "flipping" his wrists and thus adding loft with his swing. He will be unlikely to see much change in his trajectory from selecting a more tip-stiff shaft design. The golfer below has a "late" release. He almost certainly will see different trajectories with the same clubhead from using shafts with different bend profiles.

How is the "coefficient of restitution" measured?

Why is the transfer of energy from clubface to ball important?

When a golf club impacts a golf ball, the vast majority of energy lost (as much as 99 percent, according to some estimates) comes from the ball compressing and deforming against the clubface, with the balance being lost to the clubhead. Using those figures, halving the amount of energy absorbed by the clubhead would only result in about 0.5 percent more energy available to be converted to ball speed, compared with almost 50 percent more energy from halving the ball losses. One way to produce more ball speed and potentially more distance would be to re-engineer the golf ball, but they are subject to an initial velocity limit specified by the rulemakers. However, the combination of bigger driver heads and advances in material technology has allowed the clubface to be made thinner and thus flex inward at impact. While that increases the very small amount of energy absorbed by the clubhead, it greatly reduces the much greater amount of energy lost to the golf ball—and the net result is an increase in the initial ball speed.

The United States Golf Association (USGA) considered this development a threat to the integrity of the game and, in 2002, imposed an arbitrary coefficient of restitution (COR) limit of 0.83 on driving clubs with lofts of 15 degrees or less (see caption at top of page 75 for an explanation of COR). Although they initially resisted a similar decision, the Royal and Ancient (R&A) eventually followed suit. In 2006, the 0.83 COR limit was extended to apply to *all* clubs. How much has imposing a COR limit actually affected driving distance? All else remaining equal, if you have a driver clubhead speed around 95 mph (153 km/h), the distance you might expect to hit an "illegal" driver with a 0.84 COR is only a yard or two more than you would get from a "legal" driver with a COR of 0.83.

However, if you use a non-conforming driver in a qualifying competition, you may or may not be disqualified, depending on how the "conditions of competition" are phrased. There is no requirement for club manufacturers to submit clubs for COR conformity testing, but, if they do, their clubs will be placed on either the "conforming" or "non-conforming" lists of clubs maintained by the R&A and USGA, which you can see on their respective web sites. If the "conditions of competition" state that you must use a driver that is on the "conforming" list, you *must* use a driver that is on that list, but if they simply state that you may not use a driver that is on the "non-conforming" list, you are free to use any driver that is not on the "non-conforming" list, even if it is not on the "conforming" list, because drivers are presumed to be conforming unless they have been tested for conformity and failed.

Cannon and ball

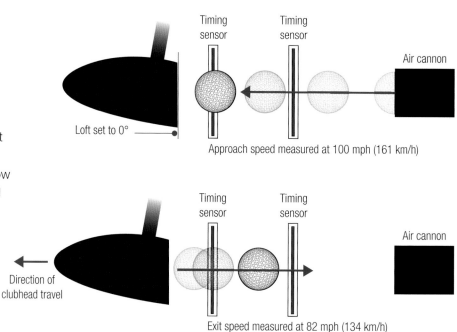

Approach speed measured at 100 mph (161 km/h)

Exit speed measured at 82 mph (134 km/h)

Bouncing balls

Solid surface

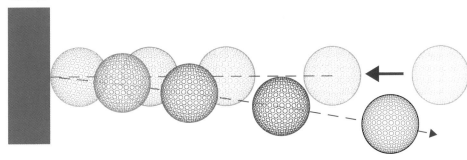

Impact speed 100 mph
(161 km/h)
Rebound speed 100 mph
(161 km/h)
COR = 1

Solid surface

Impact speed 100 mph
(161 km/h)
Rebound speed 75 mph
(121 km/h)
COR = 0.75

A simple way to appreciate the concept of COR is to imagine that a golf ball is thrown against a rigid wall at 100 mph (161 km/h) and it bounces back at 75 mph (121 km/h). The COR is therefore 75/100 = 0.75, that being the exit speed divided by the approach speed.

◀ *Coefficient of restitution*
The coefficient of restitution (COR) is a measure of the efficiency of the transfer of momentum between two colliding bodies—for example, a clubhead and a golf ball. A COR of 1.0 would mean that no energy at all is lost and a COR of 0.0 that all energy is lost. A COR of 0.830 (the limit imposed by the USGA and R&A on conforming drivers and other clubs) represents a 17 percent energy loss.

Test rig

◀ ▶ *Testing COR and CT* *The original COR test used an air cannon to fire a golf ball at a known speed at the face of the clubhead and calculated the COR based on the relative rebound velocities of the ball and the clubhead (left). In 2004, an alternative "CT test" protocol was introduced for the testing of drivers, using a portable rig in which the clubface is impacted by a steel pendulum and the "characteristic time" (the time that the pendulum is in contact with the clubface) is measured (right). The CT is directly related to the COR of the clubface. However, it has since been determined that CT only correlates with the COR (as measured by the air cannon test) for drivers, so the air cannon test has been reintroduced for clubheads other than drivers.*

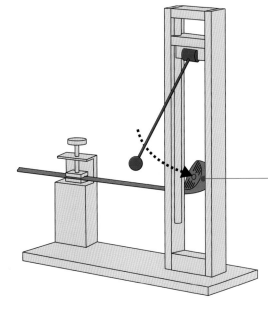

This rig was adopted by the R&A and USGA to determine whether driver heads conform to the rules on "spring-like effect," using CT (characteristic time) as a proxy for COR.

equipment: the iron

At one time, golf clubs were fashioned entirely from wood and animal bone or horn. Since about 1900, the combination of advances in manufacturing, materials technology, and clubhead design have undoubtedly changed the game and how it is played.

The early metal iron heads were often forged by local blacksmiths. Although more durable than their wooden predecessors, the lack of any perimeter weighting, their narrow soles, and sharp leading edges made them difficult to use, and they tended to destroy the expensive golf balls, or "featheries" (so-called because they were made from leather and compressed, boiled feathers). The evolution of the golf "iron" is shown below.

Steel shafts began to replace hickory in the late 1920s. Since then other developments have followed, including cavity-backed (perimeter-weighted) irons, hollow-bodied metal woods, graphite shafts, titanium wood heads, hollow-bodied hybrid iron replacements (or "rescue" clubs), and the widespread use of investment casting to make iron heads.

Until about the mid-1990s, irons were numbered from 1 to 9 and sold in sets with a pitching wedge and sand wedge. However, this has changed in recent years as the club companies have gradually decreased iron lofts—in some cases considerably. In the 1960s to 1970s, typical lofts for the 4 iron, 6 iron, 8 iron, and pitching wedge were respectively 28°, 36°, 44°, and 52°, but today you can buy iron sets with 20° 4 irons, 26° 6 irons, 34° 8 irons, and 44° pitching wedges.

▼ **Beginnings** *The basic concept of a set of clubs with different lengths and lofts to provide a progression of trajectories and distances was developed (longer shafts with lower lofts result in lower trajectories and more distance, whereas shorter shafts and higher lofts equal higher trajectories and less distance).*

▼ **1800s** *The more lofted and shorter clubs began to be made from iron (hence the name), by local blacksmiths. The combination of their design and hickory shafts made them difficult to use in comparison with modern irons.*

▼ **Late 1800s to early 1900s** *The first patent for a metal wood head was taken out in 1896 by Sir William Mills. Face scorelines (grooves) started to replace punch marks on metal iron heads. The first steel shafts were made in 1893 by Thomas Horsburgh, but were not approved for tournament use until the 1920s.*

▼ **1960s to present** *Cavity-backed iron heads produced by the investment casting (or "lost wax") process started to be produced. The first carbon fiber shafts appeared in about 1973.*

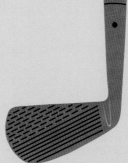
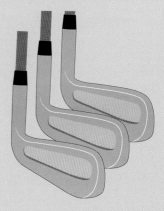

Iron anatomy

Weight distribution
A rear cavity allows weight to be moved to the heel and toe areas to reduce clubhead twisting on off-center impacts.

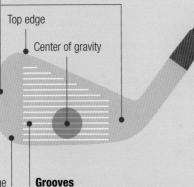

Top edge

Center of gravity

Leading edge

Grooves
Combined with face milling, grooves (or scorelines) help to promote spin (mainly from the rough or semi-rough). In 2010, the USGA and R&A changed the rules to try to limit their effectiveness.

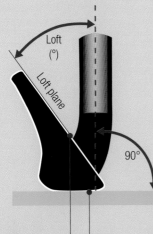

Center line

Loft (°)

Loft plane

90°

Clubface
Loft and surface roughness affect launch angle and spin. Variable-thickness face inserts made from high-strength, more "elastic" metal alloys can reduce ball speed drop-off and thus provide more consistent distance on off-center impacts..

Sole
The width and profile of the sole (front-to-back and toe-to-heel) affect the way that the club interacts with the turf.

Grip
Usually made from a rubber compound, but increasingly from synthetic materials, the grip should be customized to fit the golfer's hands and allow them to hold the club securely with minimal muscle tension in the hands, wrists, and forearms.

Shaft
Generally made from steel or carbon fiber composite, the shaft is the main determinant of the weight of the club, and, for a player with a relatively late release of the wristcock angle, can affect the launch angle, spin rate, and trajectory of the shot.

Hosel
The hosel connects the clubhead to the shaft. Additional weight can be added internally to adjust the club balance (swingweight or moment of inertia). It may also be bendable and so allow the club loft or lie to be adjusted.

Vertical to ground

Length

Lie

Horizontal to back edge of the heel

Horizontal to center

Clubhead
The loft, lie, and face angle at impact affect the launch angle, spin, initial direction, and shape (hook or draw, fade or slice) of the shot.

▲ *Iron specifications* *The iron head specifications (loft, lie, sole configuration, offset, weight, weight distribution, center of gravity location, and in some cases even its appearance at address), in conjunction with the shaft,* **and** *the final assembly specifications of the club, will all affect how well the golfer will play with it.*

What aspects of a golf club affect feel?

What is "feel"?

"Feel" is how the brain interprets all the various sensations which the nerves transmit to it when a club is swung or strikes a ball. What the golfer's conscious brain registers and how well or badly those various sensations match their personal preferences will tend to produce a subjective like or dislike for a club. Not all golfers are equally sensitive to the various features of an assembled club that affect feel. Ask one golfer to describe how a club "feels" and the response may simply be "nice" or "bad"—a good clubfitter will need to investigate further to ascertain precisely which aspect of the feel of the club the golfer likes or not. Ask another golfer and you may get instant and detailed feedback on the overall bending feel of the shaft, where in their swing and which segment (butt, mid-section, or tip) they could sense bending, the physical weight of the club, its moment of inertia or swingweight, if the impact felt "solid" or harsh, and even grip texture or size.

If a golfer feels completely "at one" with a club, they will generally swing it better, and altering certain specifications which affect the feel of a club can produce measurable differences in the way in which a golfer swings it (for example, in their swing path or the release of the club)—even if they have not consciously noticed any real change in the way it feels. Does that mean that "feel" also operates on a subconscious level, or is it simply that the laws of physics have intervened —or a combination of both?

There is little doubt that a golfer's sense of feel—or their sensitivity to particular aspects or combinations of feel—can change over time, even from day to day or hour to hour (try wielding a sledge hammer for 10 minutes and then pick up and swing your driver). "Feel" is subjective rather than absolute, which why it is a difficult term to define in a meaningful way that applies to all golfers. However, it is clear that a golfer can acquire a better sense of feel over time.

Neural feedback

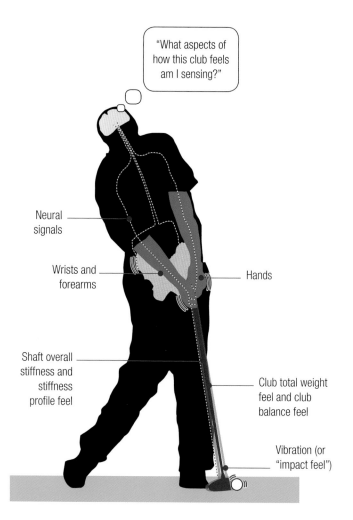

"What aspects of how this club feels am I sensing?"

Neural signals

Wrists and forearms

Hands

Shaft overall stiffness and stiffness profile feel

Club total weight feel and club balance feel

Vibration (or "impact feel")

▲ **Sensory perception** *Most of the sensory feedback which is interpreted by the brain is picked up by the nerves of the fingers, hands, wrists, and forearms, because ultimately these are the parts of the body which connect the golfer to the club.*

Feel factors

Overall stiffness and bend profile

Both the overall flex and the flex distribution of a shaft play a part in transmitting the sense of where and when the shaft is bending during the swing, which, for golfers who are sensitive to it, can have a very significant effect on their timing and the freedom with which they sense that they can swing a particular club.

Assembled moment of inertia/swingweight

The "balance" of the club (which golfers often describe in terms of the club being "head-light/head-heavy" or of releasing too early/late or too fast/slowly), is affected by the club weight and balance point in relation to the hands. The same actual moment of inertia or swingweight setting will not necessarily feel the same to all golfers—and different golfers will need different settings to provide the best feel to maintain a consistent swing tempo and time the release of the club into impact.

Club total weight

The physical weight of a club is controlled mainly by the weight of the shaft, which can vary from as little as 40 g to as much as 120 g or more. Standard, mass-produced clubs or sets of clubs tend to be built to standard lengths and swingweights, which does not allow much variation in the weights of the heads. Although weight can be measured, the same club may not actually feel heavy or light to any one golfer unless they pick up a heavier or lighter one.

Shaft flex

Clubhead impact

The sensation of the clubhead impacting the ball is affected by four things: clubhead design; where on the clubface the ball is struck; clubhead material; the flex and the flex profile of the shaft. While any of those factors can elicit comments such as "This club feels soft (or harsh, or 'dead') when I hit the ball," it is highly unlikely that impact feel can be objectively quantified in a way that is meaningful for all golfers. However, the golf shaft contributes to feel in a number of ways, which is probably why many golfers regard it as the most important component of any golf club.

▲ *How a club "feels"* *What any golfer feels when they swing a club, and when the club impacts the ball, is a composite of a number of different sensations picked up by the fingers, hands, wrists, and forearms and then interpreted by the brain. However, one particular aspect—"impact feel"—has been shown to be hugely influenced by impact sound, which the brain subjectively "translates" to impact feel.*

How does perimeter weighting reduce the effects of imperfect ball striking?

How can clubhead weight distribution help my game?

The term "sweetspot" is often misused. Strictly speaking, a ball is only truly struck on the "sweetspot" when the force vector of the center of gravity (COG) of the clubhead is aligned with the COG of the golf ball—both are simply points in space and cannot be increased in size. In practice, the term "sweetspot" has come to mean something else, namely a significantly larger area surrounding the center of the face where the ball impact can occur without the player sensing a harsh strike and without resulting in a significant loss of accuracy or distance. Therefore, the commonly accepted definition of the "sweetspot" is mainly a function of the moment of inertia (MOI) of the clubhead—so the greater that is, the larger the "sweetspot" will be.

When a ball is struck toward the toe or heel of the clubhead, the clubhead will twist about a vertical axis that passes through its COG. Shifting weight away from the COG and out toward the outer points or perimeter of the clubhead (hence the term "perimeter weighting") allows the MOI of the clubhead to be increased. The bigger and heavier the clubhead, the greater the weight that is repositioned away from the clubhead COG; and the farther away it is placed from the COG, the higher the clubhead MOI can be. Drivers—having by far the biggest heads—can have the highest MOIs. Clubheads are subject to maximum limits imposed by the rulemakers on both size (460 cc) and MOI (5900 g/cm²)—so clubheads cannot be made significantly more "forgiving" than they currently are.

Because of the practicalities of building clubs that golfers can swing effectively and comfortably, the sizes and weights of

fairway wood, hybrid, and iron heads are pretty much fixed, so they probably cannot be engineered to twist any less for a given off-center impact than now. So for all clubs—including a few high-tech, multi-material iron heads—the design emphasis has shifted to developing highly engineered, variable-thickness faces made from high-strength steel and titanium alloys which can reduce the amount of drop-off in ball speed and distance that normally occur with off-center impacts.

Clubhead designs and MOIs

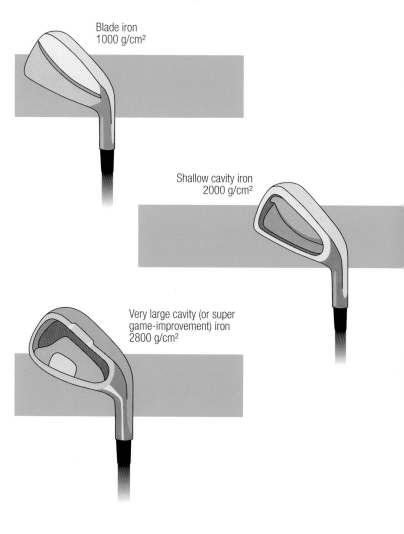

Blade iron
1000 g/cm²

Shallow cavity iron
2000 g/cm²

Very large cavity (or super game-improvement) iron
2800 g/cm²

▶ *Maximizing forgiveness* *Various designs of clubs have evolved to increase resistance to twisting and ball speed drop-off on off-center impacts. Modern methods of manufacture have allowed club designers to position more clubhead weight away from the center of gravity of the club, toward the outer edges or sides of the clubhead, thus increasing the clubhead's moment of inertia, and subsequently enlarging the sweetspot on the clubface.*

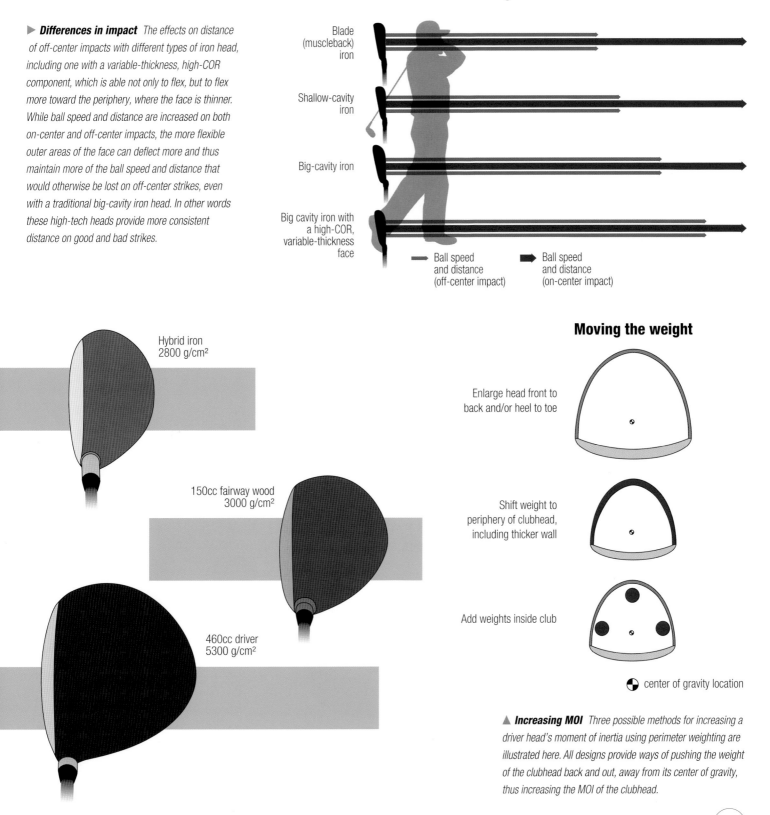

► **Differences in impact** *The effects on distance of off-center impacts with different types of iron head, including one with a variable-thickness, high-COR component, which is able not only to flex, but to flex more toward the periphery, where the face is thinner. While ball speed and distance are increased on both on-center and off-center impacts, the more flexible outer areas of the face can deflect more and thus maintain more of the ball speed and distance that would otherwise be lost on off-center strikes, even with a traditional big-cavity iron head. In other words these high-tech heads provide more consistent distance on good and bad strikes.*

Reducing the difference

Blade (muscleback) iron

Shallow-cavity iron

Big-cavity iron

Big cavity iron with a high-COR, variable-thickness face

━━ Ball speed and distance (off-center impact)

➡ Ball speed and distance (on-center impact)

Hybrid iron
2800 g/cm²

150cc fairway wood
3000 g/cm²

460cc driver
5300 g/cm²

Moving the weight

Enlarge head front to back and/or heel to toe

Shift weight to periphery of clubhead, including thicker wall

Add weights inside club

⊕ center of gravity location

▲ **Increasing MOI** *Three possible methods for increasing a driver head's moment of inertia using perimeter weighting are illustrated here. All designs provide ways of pushing the weight of the clubhead back and out, away from its center of gravity, thus increasing the MOI of the clubhead.*

Are clubhead materials and manufacturing methods significant?

Does how the clubhead is made affect my golf?

Most iron heads are, broadly speaking, made by one of two processes: investment casting or forging. Investment-cast heads are generally made from stainless steel, but can be made from a number of other metals, including carbon steel (which is then normally chrome-plated to prevent rusting), copper-based alloys, and even titanium. Forged irons are usually made by passing a single billet of soft carbon steel sequentially through several dies in a forging press, after which it is hand-ground and chrome-plated. There is a "halfway house" between the two methods (called "form forging" or "coin forging"), in which the head is first cast to its basic shape and then finished in a forging press.

There is no difference in performance (launch angle, spin, or distance) between a forged and a cast iron head of the same design. The more skilled the golfer, the less they will need a significant amount of perimeter weighting. Most forged heads tend to be targeted at this group of players, so tend to be less "forgiving" than typical cast heads, but there are some very forgiving forged irons available with large cavities milled in the backs of the heads. So-called "game improvement" designs are more likely to be cast with large cavities (to maximize perimeter weighting) and other features like wider soles, lower centers of gravity, and more offset to help improve shot outcomes for less-skilled golfers by making the clubs more "playable." However, all those things are functions of the head design, not the manufacturing method or the materials used.

Wood heads are made from a variety of materials. Drivers are almost universally made from titanium alloys these days (and are increasingly being fabricated from three to four segments), while fairway woods are still mainly made from stainless steel. Some companies have produced multi-material driver and fairway wood heads mainly to manipulate the clubhead center of gravity (for example, lowering it by making the crown of carbon composite and thus influencing launch angle and spin), rather than increasing ball speed, although combining advanced metal alloys and clever face design means that some fairway woods are now pushing up against the 0.83 COR (coefficient of restitution) limit, which previously they could not.

Casting iron heads

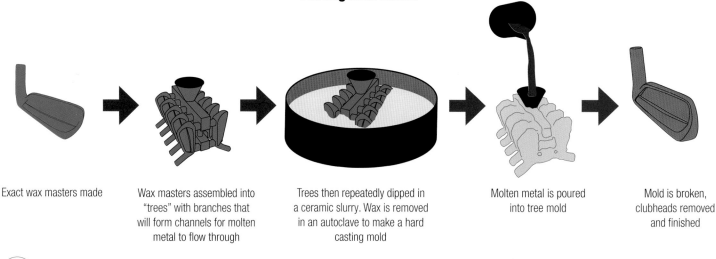

Exact wax masters made

Wax masters assembled into "trees" with branches that will form channels for molten metal to flow through

Trees then repeatedly dipped in a ceramic slurry. Wax is removed in an autoclave to make a hard casting mold

Molten metal is poured into tree mold

Mold is broken, clubheads removed and finished

Driver

Driver heads tend to be made of titanium because they can be made larger and more "forgiving" in off-center impacts.

Titanium or carbon composite crown piece.

Varying the face thickness can reduce distance loss on off-center hits.

Iron

If the face area of an iron is replaced with a separate, lighter insert, the weight saved can be moved to other areas to enhance playability and forgiveness.

If the face insert is made from a material that allows it to flex and reduces in thickness toward the edges, off-center impact performance can be improved even more.

Increasing the size (blade length) of a cavity-backed iron head will automatically increase its MOI (moment of inertia), also improving forgiveness.

▲ *Material choices* *The decision about what material and manufacturing process will be adopted for any clubhead depends on many factors: required performance and adjustability, production costs, and even what golfers expect. While it's technically possible to cast blade (muscleback) irons from stainless steel and fabricate driver heads from high-strength steel alloys that are right on the 0.83 coefficient of restitution limit, a set of blades made from anything other than forged carbon steel, or a driver that isn't made from a titanium alloy, might not be acceptable to many golfers, and thus not sell very well.*

◄ *Casting process* *A golf club consists simply of a head, a shaft, and a grip, which are mostly made by specialist foundries, shaft and grip manufacturers. They are shipped to plants owned by the club companies whose brand names they bear for assembly into finished clubs. Iron heads are made by casting or forging, with some of the newer "high tech" designs incorporating separate face pieces to enhance performance. Wood heads (not shown) are increasingly being fabricated in three or four segments which are welded together, rather than cast.*

Forging an iron head

Heated and bent steel billet placed in first die

Two or more dies progressively form head shape

Rough forged head produced for final grinding, machining and plating

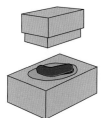
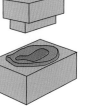
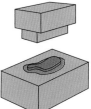

knowing the distances

Once you have selected your golf clubs to suit your swing style, and then focused on developing consistent, centered contact with the ball, the next thing to work on is understanding your shot distances. Remembering how far you hit the ball with each club—knowing your "yardages"—will help to build consistency.

You can calculate your yardage for each club, and for different levels of swing for each club, on the practice range by hitting a dozen or so balls—remember not to overhit so your shots are accurate, reproducible, and predictable—and making a note of the average distance the ball travels. You can then use your yardages on the course, in tandem with an accurate estimate of the actual distance to target. Remember to base your club selection on your average distances with each club, and, if in doubt, hit more club rather than less—course designers generally put hazards at the fronts of greens for a reason.

Tour pros benefit from the assistance of experienced caddies who provide them with detailed distance information for each shot. Most golfers, however, have to rely on other methods. Golf courses often provide basic information for golfers in the form of course maps with tee-to-hole distances, or distance markers on the course, but new technology is again coming to the rescue. Devices which use a global positioning system (GPS) offer a way of pinpointing the golfer's position, and the position of the target to within a couple of yards. Cellphone apps are available which use the phone's position and offline data. However, a cellphone is not always within signal range. Laser-rangefinder devices are also more common today, and these offer another way to estimate distances.

Carry distances will change with wind speed and direction, and the total distance— which includes the roll of the ball on impact—will depend on the condition of the fairway or rough. Once you have an idea of the distance, the choice of shot, and club—and taking the shot—are still down to you.

▶ **Heads together** *English golfer Ross Fisher consults with his caddy Adam Marrow. While Tour pros are able to call on the assistance of experienced and knowledgeable course guides, they also need to know exactly how far they will hit each shot with each club, just like the rest of us. Using this knowledge in conjunction with GPS devices, cellphone apps, laser rangefinders, and the humble course map will help take the guesswork out of each shot.*

How can the "smash factor" be increased?

How can I hit the ball better?

Understanding "smash factor" is important for several reasons. In the first place, it correlates strongly with how solidly the ball has been struck, so provides useful information about the "centeredness," "squareness," and consistency of impact. Secondly, it influences the initial ball speed that will be produced with any club at a given clubhead speed, which has implications not only with respect to the consistency with which distance can be controlled (important on approach shots), but also on the maximum distance that can be achieved with any club (important for drivers).

Smash factor is simply the initial speed of the golf ball when it leaves the clubface divided by the clubhead speed at impact. The maximum value for an on-center and square impact for a given club depends on its loft and the coefficient of restitution (COR) of the clubface at the point of impact with the ball; the highest values will be seen with low-lofted clubs with faces that are designed and made from materials that allow them to flex—drivers, fairway woods, hybrids, and, more recently, some multi-material iron heads. For a typical driver fitted with a head that is right on the COR limit allowed under the rules, a perfect on-center strike will result in an initial ball speed about 1½ times faster than the clubhead speed at impact, which gives a smash factor of 1.5.

Tour players are able consistently to hit the ball on or very close to face center with all their clubs, not only because of their level of skill, but also because their clubs are properly fitted for length, weight, dynamic balance, shaft flex, and bend profile. The lower a player's skill level and/or the less well their clubs fit them, the more their smash factors will tend to be sub-optimal; it is not uncommon, for example, to see higher-handicappers using drivers that are longer, or lighter/heavier in terms of total weight or MOI/swingweight than they can properly control with smash factors close to 1.40, or in some cases lower than that, which will certainly be costing them distance.

▼ *Factor formula* *With the appropriate equipment, it is possible to measure the necessary inputs so you can calculate the maximum "smash factor" achievable with a particular clubhead using the following formula:*

$$\text{SMASH FACTOR} = (1 + \text{COR}) \, \frac{\cos(\text{SPIN LOFT})}{1 + \frac{\text{BALL MASS}}{\text{CLUBHEAD MASS}}}$$

Where:

COR is the coefficient of restitution of the clubhead

Spin loft is the effective loft (or dynamic loft) of the clubhead at impact relative to the angle of attack

Effective loft is the loft of the clubhead at the point of impact (as measured in a gauge), plus any loft added by shaft bending, plus any loft added by the golfer's swing mechanics

Angle of attack is the vertical (up–down) angle at which the clubhead is moving at impact, relative to the horizontal; a positive angle of attack means hitting up on the ball, a negative angle of attack means hitting down on the ball

Ball mass is the actual weight of the ball in ounces (or grams)

Clubhead mass is the weight of the clubhead in ounces (or grams). It must be the actual weight of the head as fitted to the shaft to achieve the desired balance at the chosen club length, which may be quite a bit different from the nominal specified head weight.

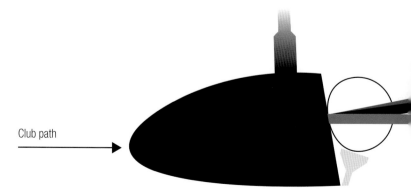

Club path

▶ Max smash The table on the right shows the average clubhead speeds, smash factors and ball speeds for the Men's PGA Tour in 2010—adapted from data collected and analyzed by the makers of the TrackMan™ launch monitor.[1] While few amateur golfers can expect to achieve the same clubhead speeds and distances as some of the best players in the world, with proper coaching and clubs that fit their strength, build, and swing profile, every golfer should be able to get close to attaining similar smash factor levels. A golfer with a 95 mph (153 km/h) clubhead speed and a smash factor of 1.40 could expect to get about an additional 10–12 yd (9–11 m) out of his driver from simply improving his smash factor to 1.47. Increasing it from 1.40 to 1.50 could possibly produce as much as 17–20 yd (16–18 m) more.

Regardless of the clubhead speed, the ball will leave the driver face at an average of ~87 percent of the spin loft at the point of impact (+/– ~2.5°). Adding or subtracting the angle of attack will give the ball launch angle relative to the horizontal.

Example: Spin loft at point of impact = 12° and 87% of 12° = 10.5°. For a level (0°) angle of attack, the launch angle will be ~10.5° (~10.5° +/–0°); for a 3° positive angle of attack the launch angle will be ~13.5° (~10.5° + 3°); for a 3° negative angle of attack, the launch angle will be ~7.5° (~10.5° –3°). The diagram below was adapted from an original graphic by TrackMan™.[2]

Dynamic (or "effective") loft is measured relative to the vertical and *includes* the angle of attack. Spin loft is measured relative to the AOA, so the two are only the same when the AOA is level (0°)

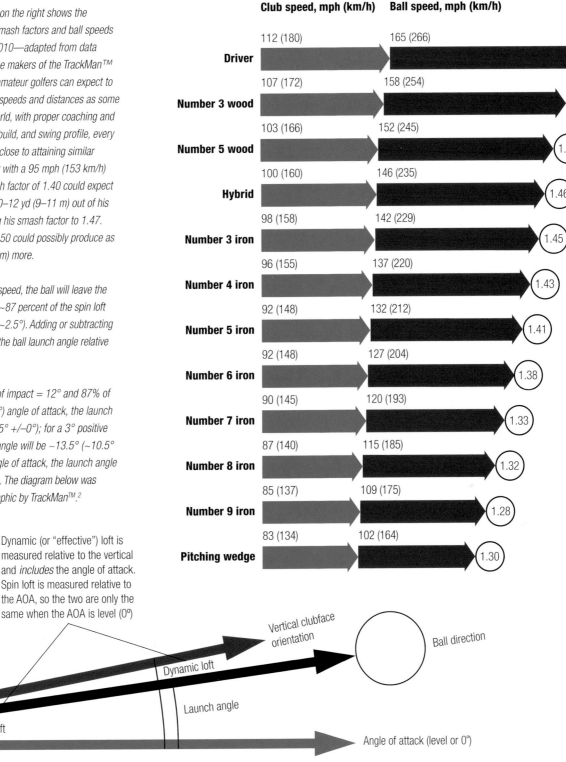

	Club speed, mph (km/h)	Ball speed, mph (km/h)	Smash factor
Driver	112 (180)	165 (266)	1.49
Number 3 wood	107 (172)	158 (254)	1.48
Number 5 wood	103 (166)	152 (245)	1.47
Hybrid	100 (160)	146 (235)	1.46
Number 3 iron	98 (158)	142 (229)	1.45
Number 4 iron	96 (155)	137 (220)	1.43
Number 5 iron	92 (148)	132 (212)	1.41
Number 6 iron	92 (148)	127 (204)	1.38
Number 7 iron	90 (145)	120 (193)	1.33
Number 8 iron	87 (140)	115 (185)	1.32
Number 9 iron	85 (137)	109 (175)	1.28
Pitching wedge	83 (134)	102 (164)	1.30

Vertical clubface orientation

Ball direction

Dynamic loft

Launch angle

Spin loft

Angle of attack (level or 0°)

Which design elements of a clubhead affect distance?

How does clubhead design help me hit the ball farther?

A perfect golf swing should result in the ball being hit on face center, but only very skilled golfers can do that consistently; most need some help from the clubhead designer to maximize distance (and accuracy/consistency) when they don't. As a result, golf clubheads incorporate a range of features to improve their playability for different golfer and swing types, helping to offset the effects of poor swing moves and ball striking. So how can clubhead design affect a golfer's game?

The clubhead center of gravity (COG) can affect distance in several ways. Moving the COG location forward or back in relation to the shaft axis can cause more or less forward shaft bending at impact, which—for a golfer with good swing fundamentals—dynamically affects the loft of the clubhead, but less for irons than for woods and hybrids. Moving it up or down influences launch angle and backspin, but the effect is different for irons than for woods and particularly drivers. By designing the face so that it can deform when it strikes the golf ball, designers have been able to increase the coefficient of restitution (COR) of the impact between clubface and ball, but only up to the current 0.83 COR limit set by the rules. That is only possible for clubheads with thin faces made from special high-strength steel or titanium alloys, and not for traditional iron heads, which are cast or forged from a single material. As a result, the current design emphasis is now on trying to maintain the COR as close as possible to 0.83 over a larger area of the face.

The loft of the clubhead, in conjunction with the shaft and the golfer's swing mechanics, directly affects the vertical angle and the spin rate at which the golf ball leaves the clubface. Both need to be optimized if maximum distance is to be achieved

Center of gravity

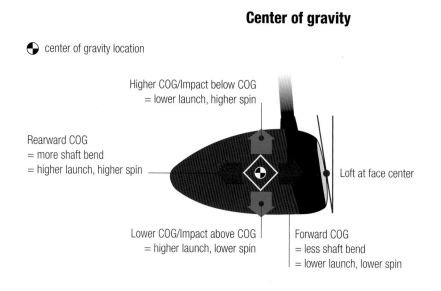

◉ center of gravity location

Higher COG/Impact below COG = lower launch, higher spin

Rearward COG = more shaft bend = higher launch, higher spin

Loft at face center

Lower COG/Impact above COG = higher launch, lower spin

Forward COG = less shaft bend = lower launch, lower spin

◀▶ **Effect of COG location** *Moving the COG location of a large, deep-faced driver head up/down or back/forward can affect ball flight and distance in several ways. If the ball is struck above the COG, the head will rotate backward, thereby reducing the amount of backspin with which the ball leaves the face (see left) through what is called Vertical Gear Effect (an impact below the COG will increase the spin). If the clubface is also made with a top-to-bottom radius (called vertical roll), the greater loft on the upper part of the face will cause the ball to launch higher than it otherwise would, and vice versa (see right). Note that vertical gear effect can be ignored with irons, because the clubhead COG is much closer to the clubface.*

with a driver. The combination of launch angle and spin rate that will result in maximum distance is not the same for all golfers; it can and does vary considerably, depending on their clubhead speed and angle of attack (AOA), which is the angle at which the clubhead is traveling at impact relative to the horizontal and which can be downward, level, or upward.

While clubhead mass can affect distance, it can effectively be ignored because—for reasons that would take more space to explain than is available here—it is, and will probably remain, relatively static.

Vertical face roll and vertical gear effect

Loft angle

Higher impact =
higher loft and launch

Head rotates backward =
backspin decrease

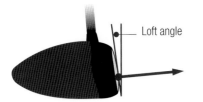

Loft angle

Lower impact =
lower loft and launch

Head rotates forward =
backspin increase

▲ **Moment of inertia** *The clubhead moment of inertia (MOI) influences distance, because it affects the amount that the head rotates closed (or open) on miss-hits toward the heel (or toe). The higher its MOI, the less it will twist, the more energy is transferred to the ball and the less it will tend to land off-line (and vice versa). However, the rulemakers have imposed a MOI limit on all clubheads of 5900 g/cm²—and only drivers, because of their shape and size (which is limited to 460 cc) will ever get close to that.*

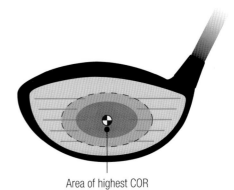

Area of highest COR

▲ **Clubhead features** *In order of importance, the main aspects of clubhead design that affect ball flight and distance for all clubs are loft, vertical clubhead COG location (left), and clubhead MOI (above). For woods (particularly drivers), front/rear COG location can play a part, as can designing the clubface to flex on impact with the ball and varying the thickness of the face component to maintain ball velocity and distance on off-center impacts (see above).*

If mastering the golf swing isn't hard enough, trying to predict how the ball will move through the air or roll on the green is not just a matter of art, but also one of science. Golf is not played in a vacuum, but rather in a constantly changing environment in which the player and equipment are exposed to a variety of conditions. By applying scientific principles to explain how environmental factors such as rain, wind, temperature, or even grass type can influence your score, Andrew Collinson and Sandy Willmott provide answers to some of the more perplexing questions about the impact the environment has on our golf performance.

chapter four

the environment

Andrew Collinson and Sandy Willmott

How does the direction of the wind affect the golf ball?

How much distance do I lose into the wind?

Hitting into a headwind or tailwind can be problematic even for the expert golfer. Aerodynamically what is important is the ball's speed relative to the air rather than to the ground. This speed is higher when there is a headwind, and the drag force on the ball increases—leading to a shorter shot. Perhaps more surprising is that the ball also flies higher because a ball with backspin experiences a lift force due to the Magnus effect (see opposite), and this force also increases as the speed of the air relative to the ball increases. The higher trajectory compounds these effects because the wind speed will typically increase with height, as interaction with the ground tends to slow the air at lower levels.

The speed of the ball relative to the air is lower when it is hit into a tailwind, and the ball will travel farther and lower as a result of smaller drag and lift forces. If the ball has sidespin, the Magnus force now points sideways rather than upward and causes the ball to fade or draw. This force will be larger, and the sideways motion more pronounced, when hitting into a headwind.

It is therefore apparent that the main problem with playing into the wind is attempting to predict the effect it will have on the ball. An expert golfer will normally try to minimize the influence that the wind has by reducing the height of their shot. This involves hitting the ball with a lower launch angle and less backspin.

Forces on the ball during flight

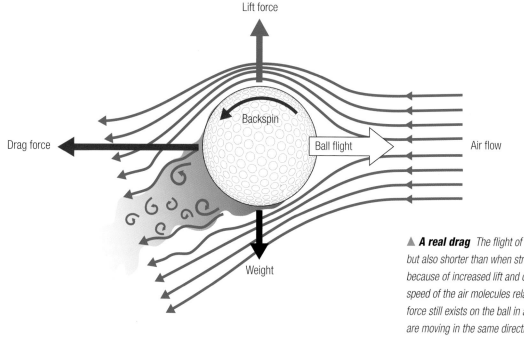

▲ **A real drag** *The flight of the ball into a headwind will be higher but also shorter than when struck into a tailwind or into still air. This is because of increased lift and drag forces, caused by the increased speed of the air molecules relative to the motion of the ball. A drag force still exists on the ball in a tailwind even though the air molecules are moving in the same direction as the ball. This is because the speed of the ball is greater than the speed of the air molecules, meaning it is still the front of the ball that contacts the air first.*

Headwinds and tailwinds

▼ **Into the wind** *The loss of distance for a shot struck into a headwind is greater than the distance gained for a shot struck with a tailwind of the same magnitude. One factor contributing to this is an increase in the wind speed relative to the ground as the height of the ball increases. Friction between the ground and the air immediately above it means that the wind speed is zero at ground level (the "no slip condition"), but increases as the air gets farther from the ground's influence.*

Shot into
a headwind

Shot into
no wind

Shot into
a tailwind

Headwind

No wind

Tailwind

The wind will also influence the roll of the ball when it lands. The higher trajectory of the shot into a headwind will cause a steeper descent of the ball, which will reduce the distance the ball rolls upon landing. With a shot into a tailwind, the distance the ball rolls will increase as the descent of the ball will be shallower.

▼ **Magnus effect** *As the ball starts to spin a thin layer of air—the boundary layer—is pulled round with it. When the ball has backspin, the air in the boundary layer is moving against the oncoming wind at the bottom of the ball and with the oncoming wind at the top of the ball. This results in an asymmetric flow with the boundary layer breaking away—or separating—from the ball's surface farther forward on the bottom side (see opposite). The wake behind the ball is deflected downward, indicating that the ball is making a downward force on the air. By Newton's Third Law of Motion the air in return must be making an upward force on the ball, which is the Magnus force.*

The effect of spin

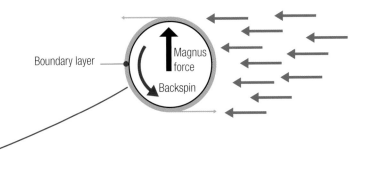

Boundary layer

Magnus
force

Backspin

How does water in the air affect the flight of the ball?

Why are my drives shorter in the rain?

It might seem logical to assume that the presence of moisture in the air, be it in the form of raindrops or water vapor, would have a negative effect on the flight of the golf ball, leading to shorter shots. A raindrop hitting the ball will indeed slow the ball down, but by how much? Actually surprisingly little because the magnitude of a raindrop's momentum—its mass multiplied by its speed—is much smaller than that of ball's momentum.

A golf ball has a mass of 1.6 oz (45.9 g) and a typical initial speed of 230 ft/s (70 m/s). If an average raindrop has a mass of 0.0003 oz (0.008 g) and travels at 15.6 ft/s (4.75 m/s) then even the worst case of a head-on collision will only reduce the speed of the ball by 0.043 ft/s (0.013 m/s) if the collision between the ball and raindrop is considered to be an inelastic one. If the raindrop's mass and speed were doubled, the ball would still lose less than 0.092 ft/s (0.028 m/s) per collision.

So even multiple collisions with raindrops during a ball's flight will not have a significant effect on its trajectory.

Increased water vapor in the air actually reduces the density of the air because water has a lower molecular weight than the nitrogen and oxygen it replaces. A reduced air density leads to a lower drag force and increased driving distance. However, rain may reduce driving distances in a range of other ways. It may reduce the firmness of the surface, reducing the roll of the ball upon landing. It will also influence a golfer's choice of clothing—waterproofs can affect a golfer's swing, reducing its length and therefore reducing clubhead speed at impact. Finally, because of the very nature of the formation of rain, the probability of wind is also increased. So although raindrops themselves may not have a meaningful effect on the ball during its flight, it is likely that many other external factors associated with rain will reduce the distance that a player hits the ball.

Moisture matters

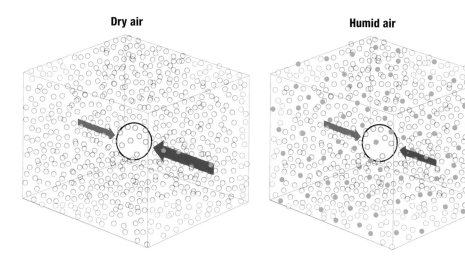

Dry air

Humid air

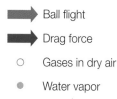

◄ **Ball flight** An increase in humidity actually increases the distance the ball will travel through the air. This is because the relative molecular weight of water vapor (H_2O, 18) is less than that of the other molecules that make up the composition of the air—mainly nitrogen (N_2, 28) and oxygen (O_2, 32)—therefore reducing the air's overall density. A lower air density results in a lower drag force (but also a lower lift force on the ball, which will counteract some of the gain from the lower drag).

➤ Ball flight

➤ Drag force

○ Gases in dry air

● Water vapor

Rain effect

Need to know

The reason the ball will travel farther in humid air can be demonstrated by the formula for drag:

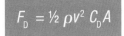

$$F_D = \tfrac{1}{2}\,\rho v^2\, C_D A$$

where F_D is the drag force, ρ is the density of the air, v is the speed of the air relative to the ball, C_D is the drag coefficient, and A is a reference area describing the size of the ball. As C_D and A will remain the same and ρ will be smaller in humid air, F_D will also be reduced for any given v.

Golf ball mass = 1.6 oz (45.9 g)
Golf ball speed = 230 ft/s (70 m/s)
Momentum = 23 lb ft/s (3.2 kg m/s)

Raindrop mass = 0.0003 oz (0.008 g)
Raindrop speed = 15.6 ft/s (4.75 m/s)
Momentum = 0.0003 lb ft/s (0.00004 kg m/s)

Raindrop

Mass = 0.0003 oz (0.008 g)

Speed = 15.6 ft/s (4.75 m/s)

Momentum = 0.0003 lb ft/s (0.00004 kg m/s)

▲▼ **Drop in the ocean** The effect that a raindrop will have on the golf ball during its flight is insignificant. The mass and speed of the ball are considerably greater than those of the average raindrop. If the mass and speed of a golf ball are 1.6 oz (45.9 g) and 230 ft/s (70 m/s) respectively and the average raindrop's mass and speed are 0.0003 oz (0.008 g) and 15.6 ft/s (4.75 m/s) respectively, then the ball will only be slowed down by 0.043 ft/s (0.013 m/s) in a head-on collision with a raindrop. Despite the fact that there may be multiple collisions during its flight with individual raindrops, the speed of the ball and therefore the distance it travels will hardly be affected.

Ball

Mass 1.6 oz (45.9 g)

Speed 230 ft/s (70 m/s) (off club)

Momentum 23 lb ft/s (3.2 kg m/s)

95

How does the air temperature affect the distance of a drive?

Why do driving distances increase in warm weather?

The majority of golfers will know that the ball travels farther on a hotter day, but why does this happen? Changes to the ball's initial speed and to the aerodynamic forces it experiences are involved. Firstly, the impact between the clubface and ball will become more "elastic" at higher temperatures, leading to less energy loss and a higher ball speed. Secondly, the molecules in the air have more kinetic energy at a higher temperature and this results in a lower air density—there are fewer molecules in a given volume. Lower air density leads to less drag on the ball, and an increase in drive distance. However, the lower air density at higher air temperatures will also have the less desirable effect of reducing the lift force on the ball, which would tend to decrease the distance the ball travels (in turn reducing the gains in distance arising from the lower drag). Generally, the ball will travel farther through the air when the temperature of the air is increased, but calculating the precise effect that changes in temperature will have on driving distance can be complicated as this depends on the launch speed, launch angle, and spin rate—which will vary for each player.

If you normally play where the air temperature is fairly consistent then accounting for changes in air temperature may not be necessary. However, changes in air temperature from day to day or even hour to hour will affect driving distances and club selection on approach shots: teeing off during the warmest part of the day—which is usually in mid- to late afternoon—may add a few yards to your driving distance.

▼ *Energy loss* *During contact between the clubface and the ball, the ball compresses and then regains its original shape. The energy returned during the expansion is always less than the energy required for the compression; some energy is lost as heat and sound. The coefficient of restitution for the collision is a measure of the ratio between the speed at which the ball and club separate after impact and the speed at which they approach (i.e. the clubhead speed), and it increases with a rise in ball temperature.*

Making an impact

Drag force

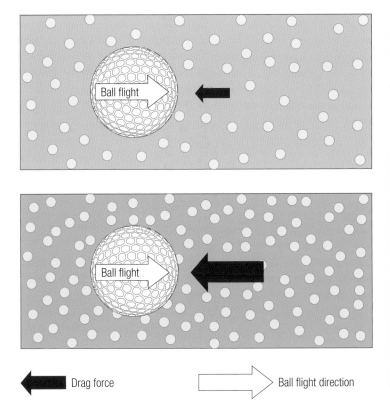

◀ **Reduced drag** *An increase in air temperature has the effect of reducing the density of the air (its mass per unit volume). This reduction in air density therefore reduces the drag force (indicated with red arrows) experienced by the ball during its flight, increasing the distance it will travel.*

■ Drag force   ⬜ Ball flight direction

Need to know

The relationship between air density, pressure, and temperature can be represented using the formula:

$$\rho = \frac{P}{RT}$$

where ρ is the density of the air, P is the air pressure, R is the gas constant for dry air, and T is the absolute temperature (i.e. in kelvins rather than degrees Fahrenheit or Celsius) of the air. So if P and R stay constant and T increases, then ρ (the air density) will decrease.

▶ **Increased distance** *An increase in air temperature will typically have the effect of increasing the distance that a golf ball travels through the air—although the exact magnitude of this effect will depend upon the specific combination of launch speed, angle, and spin rate. The graphs show the effect of varying air temperature for three simulated drives[1] with different combinations of launch speed, angle, and spin rate.*

—— Simulated drive—initial ball speed 183 mph (295 km/h), launch angle 13°, backspin 2140 rpm

—— Simulated drive—initial ball speed 159 mph (256 km/h), launch angle 13°, backspin 2584 rpm

—— Simulated drive—initial ball speed 162 mph (261 km/h), launch angle 9°, backspin 2475 rpm

Warming up

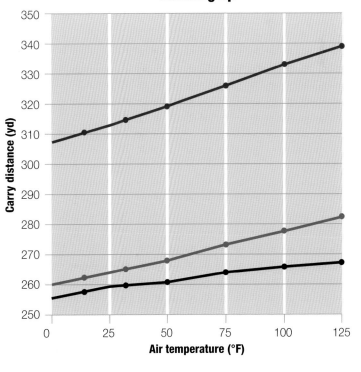

How does the landscape affect wind on the golf course?

Why does the direction of the wind change from hole to hole?

Judging the speed and direction of the wind on the golf course can be difficult even for the most expert of golfers. Wind speed typically increases the farther you are from the surface of the Earth—as a result of the reduced effect of friction with the ground. Closer to the surface, the natural landscape and terrain will change both the speed and direction of the wind. Trees on the course will further affect these properties of the wind in a complicated manner that depends on factors such as their number, height, shape, and spacing.

It's little wonder that golfers find it very difficult to judge the wind on holes such as the 12th at Augusta National. Anecdotal evidence from professional golfers suggests that the flag stick can be moving in one direction on the 12th green, and yet in a completely different direction on the 11th green—a mere 50 yd (46 m) away. Attempting to judge the wind from the flag stick or by throwing grass into the air on a tree-lined hole becomes almost redundant as this will not indicate the speed and direction of the wind above the tree line. Analyzing the movement of the clouds is a more valid method of determining the wind as its speed and direction will normally be more consistent above the tree line.

A golfer may be forgiven for thinking that judging the wind on a links course should be a lot easier than on a parkland course. Although there may be fewer trees on a links course, the natural terrain on this type of golf course may also influence the speed and direction of the wind. The sand dunes found on links courses will cause the speed of the wind to be greater on top of a dune than in front it.[1] However, there will also be a turbulent region directly behind the dune. Attempting to predict the speed and direction of the wind in this situation again becomes very difficult. Even the grandstands found at golf events will cause the speed and direction of the wind to change, funneling the wind through any gaps between them as in a tunnel. This effect may, of course, be largest at the most important—and thus popular—holes.

Gone with the wind

Height above ground

Wind speed behind trees

▶ *Factors influencing the wind* *The speed and direction of the wind are influenced by the landscape, and trees are often a major component of the landscape on a golf course. The presence of trees will normally have the effect of slowing down the wind, up to and above the height of the trees, although the exact nature of their influence on the wind can be hard to predict. The shape, number, and spacing of a given set of trees will all influence the wind speed profile behind them.*

Grandstand effect

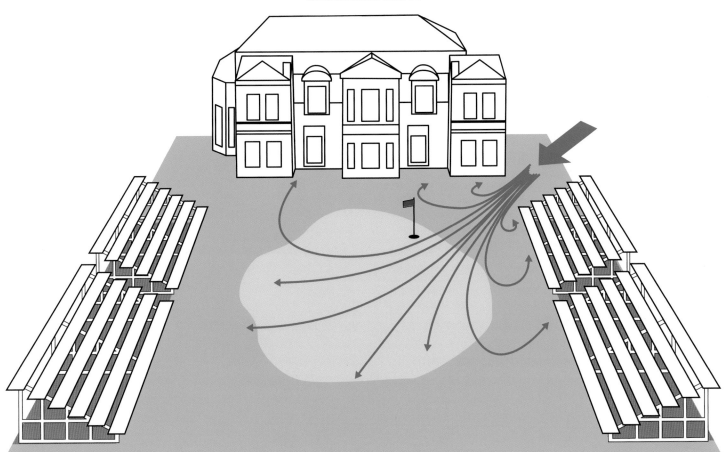

▲ **Turbulent air flow** The complexity of the air flow on a green surrounded by grandstands means it will be very difficult for a golfer to judge the wind's effect on a shot into that green. There are multiple factors which will change both the local speed and the direction of the wind. The speed will increase as the wind is channeled between two grandstands. The speed will drop again as the space opens out over the green, but the flow pattern may become very complicated with localized vortices. Moreover, this pattern will be constantly changing as the direction and strength of the incoming wind varies.

Over the hills

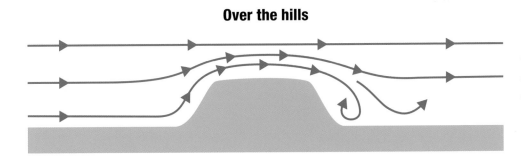

◀ **Hills and ridges** The speed of the wind can also be affected by hills and ridges on the golf course, with the air being accelerated as it is forced up and over a ridge. On the downwind side of the ridge the airflow will be more turbulent, introducing greater variability into the wind direction and speed.

equipment: the ball

There is an extraordinary amount of theoretical physics underlying the action of the humble golf ball. The design of the ball, however, came about as a result of the observations of nineteenth-century golfers rather than theories about putting feathers inside a leather casing. From around 1850, cheaper balls began to be used made from gutta-percha—a natural latex made from the sap of varieties of trees in southeast Asia. It was noticed that these balls flew farther when the smooth surface had become roughened by play, so players began deliberately to roughen the surface of their "gutties." It was not long before manufacturers were producing balls with rough surfaces, leading to the dimpled balls of today. But it was a long time before the theory caught up with the practice, and offered an explanation for why the dimpled ball went farther.

A ball traveling through the air will create a high-pressure area on its front side, with air flowing smoothly over its contours. But the air closest to the ball's surface—the boundary layer—separates from that surface farther back around the ball, generating a turbulent wake where the pressure is lower. The pressure difference across the ball causes drag. Dimples on the ball introduce turbulence into the boundary layer itself, which actually helps to delay its separation until a point farther downstream on the ball's rear side. This decreases the width of the wake and, therefore, the pressure differential across the ball. As a result, a dimpled ball has about half the drag of a smooth ball.

There is another twist that explains a further benefit of dimples— one that involves the effect of an additional force created by a ball with spin. A smooth ball with backspin generates lift by distorting the airflow so that the ball acts like an airplane's wing. The spinning action results in the air pressure below the ball being higher than the air pressure above it; this imbalance creates an upward force on the ball.[1] This phenomenon is enhanced with a dimpled ball, and the result is an increased lift force, as shown in the graph opposite.

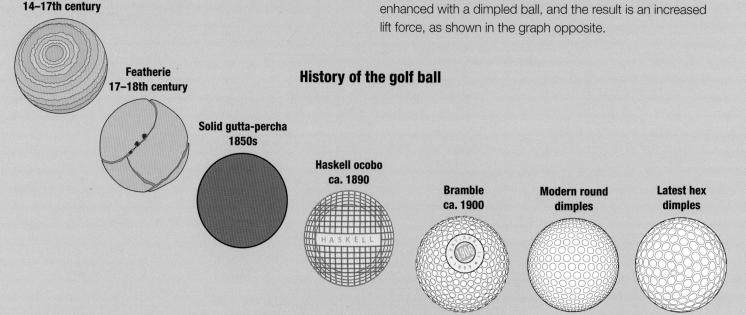

Wood
14–17th century

Featherie
17–18th century

Solid gutta-percha
1850s

History of the golf ball

Haskell ocobo
ca. 1890

HASKELL

Bramble
ca. 1900

Modern round
dimples

Latest hex
dimples

Ball flight

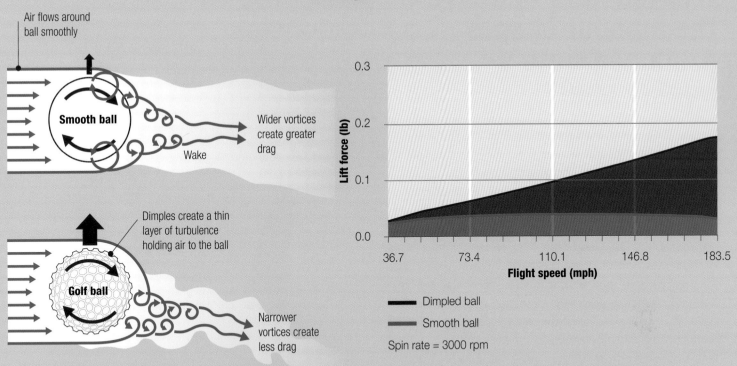

Air flows around ball smoothly

Smooth ball

Wider vortices create greater drag

Wake

Dimples create a thin layer of turbulence holding air to the ball

Golf ball

Narrower vortices create less drag

0.3

0.2

Lift force (lb)

0.1

0.0

36.7   73.4   110.1   146.8   183.5

Flight speed (mph)

— Dimpled ball
— Smooth ball

Spin rate = 3000 rpm

▲ **Point of separation** *Dimples on the ball's surface cause localized turbulence in the boundary layer next to the surface. This helps to stabilize the boundary layer, and the point at which it separates from the surface of the ball is located farther back around the surface of the ball than it would be for a smooth ball.[2]*

▲ **Lift forces** *Measurements carried out in a wind tunnel[3–5] have shown that, at identical spin rates (a typical value for the first part of a drive), the lift forces generated on a smooth ball and on a dimpled golf ball are different. As illustrated in the graph, the smooth ball doesn't create as much lift as the dimpled one, but it does create some—equivalent to about a third to a half of its own weight for much of the speed range. So it is really the spin that creates the lift—a non-spinning dimpled ball generates no lift—but dimples intensify this effect, helping to optimize a golf ball's aerodynamic performance.*

Types of golf ball

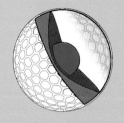

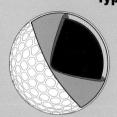

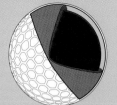

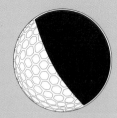

▲ **Liquid center** *A small rubber or plastic core is filled with liquid, then rubber windings are tightly wrapped around the core. Coverings such as balata, urethane, or elastomer offer higher spin rates and better hold on the greens.*

▲ **Multi-layered** *A solid rubber or plastic core is surrounded by a urethane-based layer. This in turn is encased in a cover material, which varies from brand to brand. Core size and stiffness dictate the ball's firmness, durability, distance, and spin.*

▲ **Wound core** *A large solid rubber or plastic core is surrounded by a thin layer of rubber windings tightly wrapped around it. Wound-core balls offer golfers more spin, but they lack the durability and distance of other types of balls.*

▲ **Solid core** *Constructed with a solid rubber core, surrounded by single layer, the ball is typically covered with urethanes or ionomers. This type of ball is known for its durability, distance, and reduced spin due to its dense construction.*

What factors affect the amount of spin produced from a bunker shot?

Why does my ball not check from the bunker?

The ability to control the distance that the ball will roll when playing from a greenside bunker can be a valuable asset in a player's short game, and being able to give the ball an appropriate amount of backspin is vital. A parallel, frictional force exerted by the clubface on the ball during impact generates the backspin, and this is helped by the grooves on the clubface. When hitting from a bunker both this parallel force and the normal force become attenuated as they pass through the sand, and the resultant ball speed and backspin are reduced. However, having a very thin layer of sand between the ball and clubface may actually increase the amount of backspin produced, because the sand may act like a strip of sandpaper on the clubface. The maximum amount of friction that can be produced is proportional to the coefficient of friction, which would typically be about 0.1 at impact with no sand but can be 1 or more for sandpaper.

The type of sand in a bunker can affect the depth to which the ball will plug. Analysis of the sand particles at four famous golf clubs showed that the ball was more likely to plug in the bunkers at Royal St. Davids in Wales than in the bunkers at St. Andrews in Scotland, due to the smaller average size of the sand particles at the latter course.[1,2] The ball was also less likely to plug in the bunkers at Killarney Golf Club and Moortown as their bunkers contained a greater variation of sand particles.[1,2]

So, the type and condition of the sand will affect the lie of the ball in the bunker. Understanding how such variables influence the spin rate of a shot from a green-side bunker should improve a player's short game ability. Hence, if a player is in a plugged lie they will know that the increased amount of sand between the clubhead and the ball will not only reduce the clubhead speed but also the ball speed and amount of backspin generated.

Clubface forces

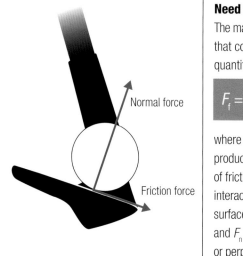

> **Need to know**
> The maximum amount of friction that could be produced can be quantified using the formula:
>
> $$F_f = \mu F_n$$
>
> where F_f is the frictional force produced, μ is the coefficient of friction (describing the interaction between the two surfaces that are in contact), and F_n represents the normal, or perpendicular, contact force.

▲ **Forces** *The clubface exerts two forces on the ball during impact. Firstly, a normal force acts at 90 degrees to the clubface. Secondly, as the ball tries to move up the clubface, a frictional force is generated to oppose this motion. The frictional force generates a torque about the center of the ball and results in the ball having backspin.*

Plugging

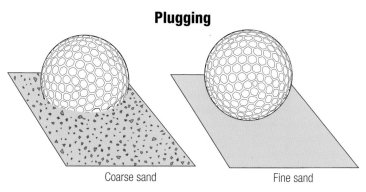

Coarse sand   Fine sand

▲ **Sand consistency** *Research has shown that raking the bunker actually increases the plugging depth of the ball. Increased moisture content, however, will reduce the depth at which the ball will plug.[2] Plugging depth increases when the sand is coarse, the sand is uniform, and/or when the sand contains a large number of spherical particles.[2]*

Hitting out of sand

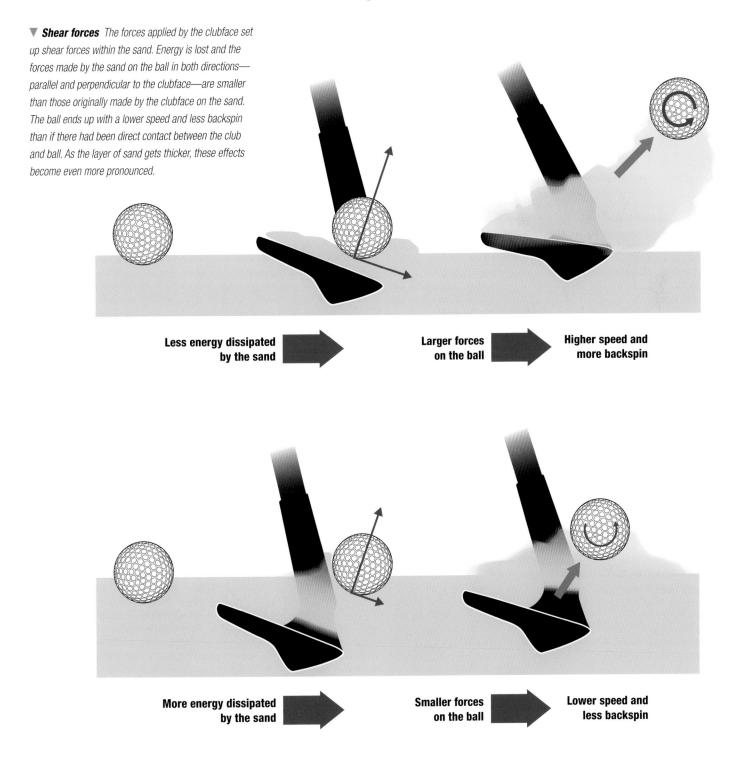

▼ **Shear forces** *The forces applied by the clubface set up shear forces within the sand. Energy is lost and the forces made by the sand on the ball in both directions—parallel and perpendicular to the clubface—are smaller than those originally made by the clubface on the sand. The ball ends up with a lower speed and less backspin than if there had been direct contact between the club and ball. As the layer of sand gets thicker, these effects become even more pronounced.*

Less energy dissipated by the sand ➡ Larger forces on the ball ➡ Higher speed and more backspin

More energy dissipated by the sand ➡ Smaller forces on the ball ➡ Lower speed and less backspin

How does the lie affect the spin on the ball?

Why doesn't the ball have backspin from the rough?

Predicting the way that the ball will fly and behave on landing when struck from the rough can be difficult, even for the most expert golfer. This is because the golfer cannot be sure of the effect that the rough grass will have on both the normal contact and frictional forces at impact. A large amount of grass will significantly reduce the normal contact force and the frictional force—with the latter decreasing the backspin produced. A study has shown that, for a 5 iron, the amount of backspin produced from the rough was on average 36 percent lower than for a shot from the fairway.[1] The reduced backspin will cause the ball to travel farther on landing although the reduced normal force at club/ball impact will have caused the launch speed and carry distance to be reduced.

It can be surprisingly difficult, even for the expert golfer, to judge the distance of a shot when there are only a few blades of grass between the clubhead and ball at impact. Despite the normal contact force remaining similar to what they may expect from a shot from the fairway, the reduced frictional force will cause the ball to travel farther on landing (because a reduced frictional force will result in less backspin). This reduction in backspin only occurs for the shorter ironed clubs.[2] So, theoretically, striking a long iron from the light rough should yield the same results as striking a shot from the fairway—as a long iron club will create less backspin compared with a short iron, even from a fairway lie.

Interestingly, many amateur golfers actually prefer to strike the ball from the semi-rough rather than from the fairway.[3] The exact reason for this remains unclear; however, the increased margin for error that this lie offers—often the ball will be sitting on top of the semi-rough—may increase the chances that an amateur has for correct ball contact (akin to striking the ball from a tee). Conversely, a professional golfer will always prefer to strike the ball from a "tight" lie found from the fairway—as they are more proficient at making correct contact with the ball compared with amateurs, and this type of lie enables them to better predict the amount of backspin produced.

A good lie?

▶ *Semi-rough strikes Research has shown that many amateur golfers actually prefer to play an iron shot from the semi-rough as opposed to a fairway lie.[3] Further analysis reveals that amateur golfers also launch the ball higher and produce more backspin from a semi-rough lie compared with a fairway lie. As well as preferring to hit the ball from a semi-rough lie, amateurs also perceived themselves as striking the ball better from this type of lie compared with a fairway lie.[3]*

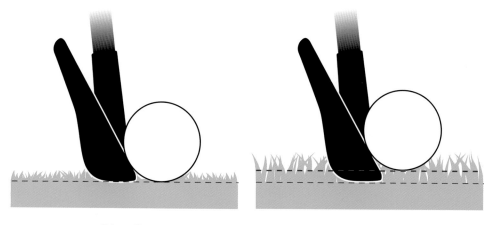

Fairway lie   Semi-rough lie

Rough forces

◀ **Thick rough** When the ball is struck from the rough, both forces on the ball at impact—the normal contact force and the friction force—are altered. When there is a large quantity of grass between the clubhead and the ball, the contact force will be reduced, meaning the ball will not fly as far in the air. The frictional force and the associated torque about the center of the ball are also reduced, and the backspin produced will be less than for a shot from the fairway. Consequently, the ball will fly lower and also run farther than when the ball is struck from the fairway.

▶ **Short rough** When there is only a small amount of grass between the clubhead and the ball at impact, the effect on the normal contact force will be negligible. However, the grass will reduce the frictional force significantly, reducing the backspin produced—although this only occurs for the shorter iron clubs. The flight of the ball will therefore be similar to that from the fairway, although slightly lower, but the ball will run farther on landing.

Types of grass

▶ **Grass differences** The type of grass found in the rough will also affect the normal contact force produced at impact. Bermuda grass has on average a greater volume per leaf than Poa annua grass. Therefore the normal contact force will usually be less at impact when striking from Bermuda rough than from Poa annua rough—meaning the ball will typically fly a shorter distance from rough that is Bermuda than from Poa annua.

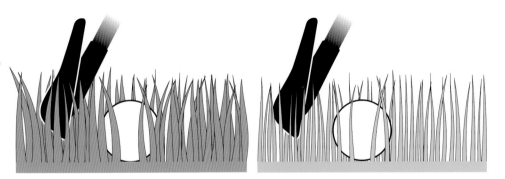

Bermuda grass

Poa annua grass

What factors affect the bounce of the ball?

How can I predict the bounce of the ball on the green?

The bounce of a ball when it lands on the green or the fairway will be influenced by two forces made on it by the ground: the normal contact force, which is perpendicular to the surface of the ground, and the frictional force, which acts parallel to the surface. Factors such as the angle of ball descent, the slope of the ground, and whether the green is elevated or depressed, will influence how the ball reacts on the surface on its first bounce.

The angle at which the ball descends will influence both the frictional and normal contact forces on the ball at ground impact. A steeper angle of descent will increase the normal force on the ball, increasing the steepness of its rebound angle and decreasing the run on the ball. The angle of descent may also cause the type of friction that is applied to the ball to differ. At moderate to large angles between the path of the ball and orientation of the ground the ball will be subject to static friction from the ground at impact. If this angle gets too small, the ball may slide during its contact with the ground and the friction will be of the sliding (or kinetic) type. Factors influencing the angle of descent of the ball include the clubhead speed at impact and the type of loft on the club: greater clubhead speed and a more lofted club will cause the descent of the ball to be steeper.

The angle of the surface will also influence the rebound angle of the ball: the ball will rebound more vertically if the slope is tilted toward the golfer, and more horizontally if the slope is tilted away from the golfer. An approach shot to an elevated green will carry a shorter distance than one to a depressed green because the ball is in the air a shorter time, but it will bounce a greater distance horizontally because it will impact at a shallower angle. These effects will be exaggerated for shots hit on a lower trajectory compared with those hit on a higher trajectory.

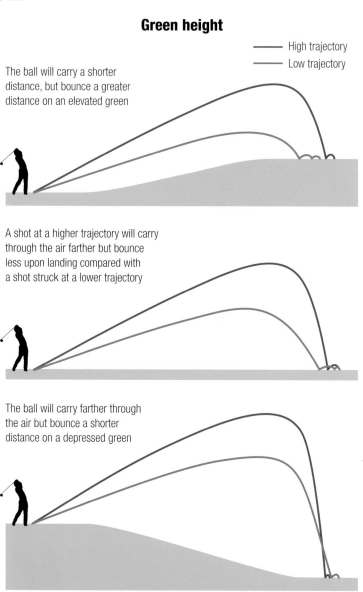

Green height

—— High trajectory
—— Low trajectory

The ball will carry a shorter distance, but bounce a greater distance on an elevated green

A shot at a higher trajectory will carry through the air farther but bounce less upon landing compared with a shot struck at a lower trajectory

The ball will carry farther through the air but bounce a shorter distance on a depressed green

▲ *Shot distance* *The distance that the ball flies and bounces/rolls on landing will be affected when hitting shots to elevated and depressed greens. On average, for a 10-yd (9-m) depression, a change to the next club with greater loft is required whereas for a 10-yd elevated green, a change to the next club with less loft is required.[1] The influence of green height varies with the trajectory of the shot. The effect of relative green height on flight and bounce/roll distance will be greater for a shot with a shallow descent angle than for one with a steeper descent angle.*

Flight and bounce

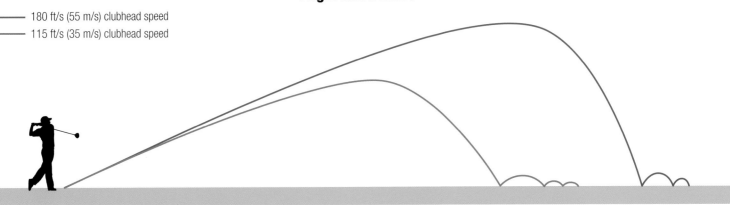

— 180 ft/s (55 m/s) clubhead speed
— 115 ft/s (35 m/s) clubhead speed

▲ ***Drive speed and trajectory*** *The speed of the clubhead at impact will influence the angle at which a golf drive descends to the ground. For a drive struck at 115 ft/s (35 m/s) the ball descent angle will be shallower than for a drive struck at 180 ft/s (55 m/s).[2] This change in the angle at which the ball impacts the ground contributes to an increased roll of the ball for a drive hit with a slower clubhead speed. Ball impact angle also changes when different iron shots are struck.[2] The impact angles of shorter irons are greater than for longer irons—as short irons fly shorter but higher and descend at a steeper angle. This results in the roll of the ball after ground contact being less for the shorter irons than for the longer irons.[2]*

▼ ***Impact angle*** *The angle of bounce of the ball on the green will be affected when there is a slope. When the ball lands on a green that slopes toward the golfer, the bounce of the ball will be more vertical than for a green with a level surface. Alternatively, when the ball lands on a green that slopes away from the golfer, the bounce of the ball will be more horizontal than for a level green. This will cause the ball to stop sooner on a green that slopes toward the golfer and travel farther on a green that slopes away from the golfer, even before the differing effects of gravity in the two situations are considered. However, the relationship between the angle of the slope and the angle at which the ball bounces is not a simple one: as the ball deforms the surface slightly it receives a contact force from the ground that is distributed over a curved contact area.[2] Predicting the bounce angle is further complicated by the fact that the ball will deform the surface to a greater extent when landing on an uphill slope than on a downhill one.*

Bounce on slopes

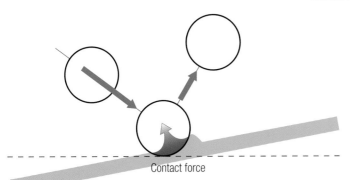

Contact force

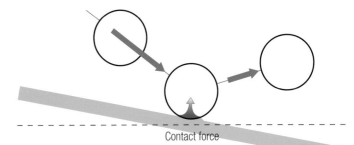

Contact force

 # considering the conditions

As we have read, cold, wet, and windy conditions can affect the flight and travel of a golf ball. They can also affect the golfer: low temperatures can make muscles tight and more prone to injury, and rain and wind will impact on a player's concentration and mental endurance.

Exposure to cold environments provides significant physiological and psychological challenges for the body.[1] In addition to the potentially harmful consequences of such conditions on core body temperature, fatigue, dehydration, and poor peripheral circulation can ensue.[2] All these factors can dramatically affect a golfer's performance on top of extreme weather conditions, such as buffeting by high winds.

The wind chill index illustrates the cooling effect of wind. Even when air temperatures are high, a cooling breeze can lower body temperatures significantly. Air currents on a windy day magnify heat loss as the warm insulating air layer that surrounds the body is continually exchanged with colder ambient air. For example, on an exposed tee, the ambient temperature of 40°F (4°C) combined with a 20 mph (32 km/h) wind will produce a wind chill temperature of 32°F (0°C). Walking at a quick pace (5 mph or 8 km/h) into the breeze, the wind speed increases to 25 mph (37 km/h), lowering the wind chill temperature further to below freezing.[3]

It is therefore important to consider ways of combating the effects of thermal stress on the body by wearing clothing made from microfiber fabrics, and other materials such as Gore-Tex®, which keep the body warm and also allow freedom of movement. Well-placed vents can also be unzipped to allow air to circulate when the body temperature rises. A lot of heat is lost through your head and neck, so wearing a hat is important. Finally, by ensuring you take on adequate nutrition, and plenty of fluids, rounds of golf in the wet, windy rain can be as productive as summer sessions—if you are motivated enough to go out in the cold weather!

▶ *Rain man* *Playing a round of golf in the rain, wind, and cold can be exhilarating, but it can quickly become a test of endurance, particularly if you are not wearing the right clothes. Here Gary Wolstenholme wears layers, including a comfortable, flexible, waterproof, and breathable jacket, and non-slip rain gloves.*

Do the grass and subsoil of a green affect backspin? → Can I stop the ball on the green?

Amateur golfers are often amazed when they see professionals land iron shots that check suddenly on the green. This amazement will then be followed by the question, "Why can't I do that?" The most likely explanation can be found in the difference between how professionals and amateurs strike the ball—professionals are able to produce more backspin. However, the greens that most amateurs regularly play on may also reduce their chances of stopping the ball quickly.

Research has shown that greens with a higher percentage of sand, together with a lower percentage of moisture, in their subsoil will allow the ball to retain more of its backspin after it makes initial contact with the green.[1] Sandy subsoil and low moisture content of a green go hand in hand—a sandy subsoil allows more water to be drained than a clay subsoil would, for instance. Drier green conditions reduce the depth of the pitchmark when the ball makes initial contact with the green, thus enabling the ball to retain backspin. Yet the reduced depth of the pitch mark also has its drawbacks, primarily increasing the distance the ball travels on its first bounce.

The greens found on links and heathland courses facilitate the retention of backspin. The subsoils for both types comprise mainly sand, often due to their proximity to the sea. The greens on moorland and parkland courses contain lower levels of sand in their subsoil. Consequently, the average moisture content of greens on links and heathland courses is lower compared with moorland and parkland courses.[2] So an approach shot will stop on an inland course primarily due to the steep rebound angle created by the deep pitch mark, whereas an approach shot onto

Deep pitch mark

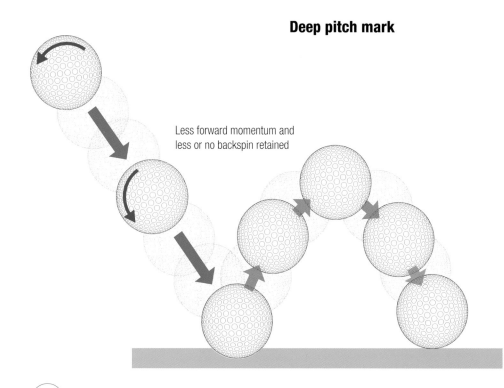

Less forward momentum and less or no backspin retained

◄ *High moisture/*Poa annua *content* *If the ball lands on a green with a high moisture and* Poa annua *grass content, then it is likely to create a deep pitch mark. This deep pitch mark has the effect of rebounding the ball at a steeper angle on its first bounce, reducing the ball's forward momentum, and causing it to stop more quickly.[1] It also removes any backspin that was originally on the ball, meaning the ball will not spin back on the green.*

► *Low moisture/*Poa annua *content* *A reduced amount of moisture in the green's subsoil and* Poa annua *on its surface will reduce the size of the ball's pitch mark upon green impact. This will cause the ball to bounce at a shallower angle and retain more of its forward momentum. However, reduced depth of the original pitch mark also enables the ball to retain much of its backspin—indeed, if the ball had enough backspin, it would spin back on the second bounce, but this is very rarely achieved.[1]*

Grass and subsoil content

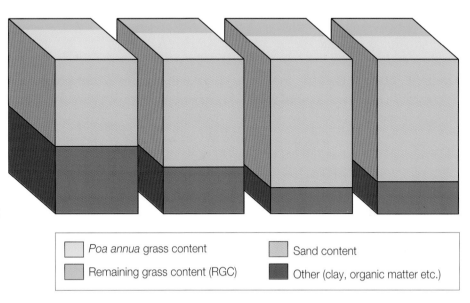

Moorland
Poa 71%, RGC 29%
Sand 58%, Other 42%

Parkland
Poa 63%, RGC 37%
Sand 72%, Other 28%

Heathland
Poa 58%, RGC 42%
Sand 85%, Other 15%

Links
Poa 58%, RGC 42%
Sand 82%, Other 18%

a seaside course will stop only if there is enough backspin on the ball. So it could be argued that the capacity to stop the ball on a links or heathland green is the true test of a player's ball-striking ability. On a links course, only professionals and players who strike the ball correctly and with enough clubhead speed will be able to generate the required amount of backspin to stop the ball quickly. But while they can stop the ball, even professionals will only very rarely be able to make a ball spin back on a links green. This is because the firmness of links greens reduces the rebound angle of the ball—leaving the ball with more forward momentum than the backspin can counteract.

	Poa annua grass content		Sand content
	Remaining grass content (RGC)		Other (clay, organic matter etc.)

▲ ***Green, green grass*** *The subsoil and grass content of greens can vary depending on the location of the golf course. Greens on links and heathland courses will normally contain a lower percentage of* Poa annua *grass but a higher percentage of sand in their subsoils compared with moorland and parkland courses.[2,3]*

Shallow pitch mark

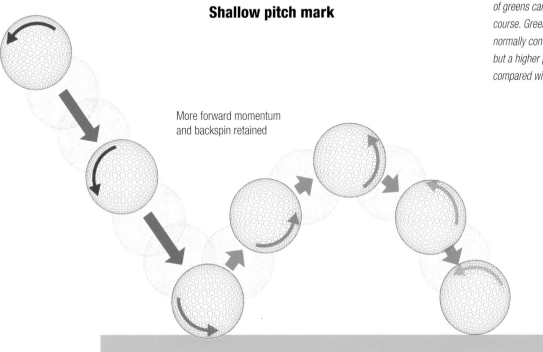

More forward momentum
and backspin retained

What forces affect the roll of the ball on the green?

Why does the ball run farther on some greens than on others?

The forces exerted on a rolling ball are actually quite complex. The translation and rotation of the ball will be influenced by gravity, the normal force, kinetic friction and/or the rolling resistance. Typically, once the golf ball leaves the putter face it will initially translate (or skid) across the surface. During this phase it will be subjected to kinetic or sliding friction which will oppose the motion of the ball. This will slow the center of mass and also create a torque, or turning force, about the center of mass, and this in turn will start the ball rolling. The ball will be considered in a state of pure roll when the contact point on the ball is moving backward relative to the ball's center of mass at the same speed as the center is moving forward.

On a sloping green, gravity will cause a downhill putt to roll farther and an uphill putt to roll shorter. The effect of gravity on the putt will be dictated by the steepness of the slope. Research has also suggested that downhill putts are more likely to stay on line than uphill putts and that this is accentuated on faster greens[1]—although, because of the effect of the slope, the ball is likely to travel farther past the hole if the putt is missed when putting downhill.

Other variables that can influence the rolling distance of the ball on the green include the mowing height of the grass, the type of grass on the green, and the frequency with which the green is irrigated. A greater mowing height, a coarse grass type, and a higher moisture level will all increase the frictional force exerted on the ball, which in turn will reduce its rolling distance.

▼ *Gravity and slope* A ball will roll farther on a downhill slope than on level ground because the ball's weight, W, contains a component W_p that is parallel to the direction of the slope and tends to accelerate the ball downhill (a role in which it is opposed by the frictional force F_f). The magnitude of W_p can be calculated using the formula:

$$W_p = W \sin\theta$$

where θ is the angle of the slope relative to the horizontal. The larger this angle the greater the value of W_p and its acceleratory effect. If the ball is instead rolling **up** a slope then W_p and F_f both point down the slope, acting together to decelerate the ball.

Effect of gravity on the ball

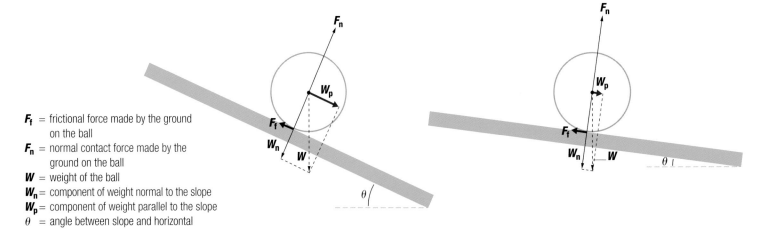

F_f = frictional force made by the ground on the ball
F_n = normal contact force made by the ground on the ball
W = weight of the ball
W_n = component of weight normal to the slope
W_p = component of weight parallel to the slope
θ = angle between slope and horizontal

Spring

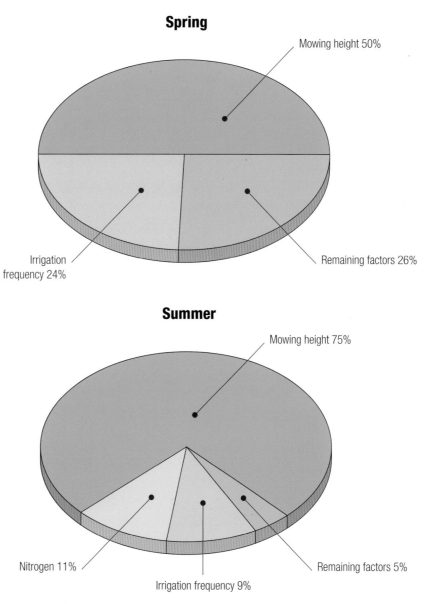

Mowing height 50%

Irrigation frequency 24%

Remaining factors 26%

Friction from grass Course management and environmental factors will also influence the distance the ball will roll on the green.[2] The most significant of these is the mowing height of the grass, as a lower mowing height reduces the friction acting on the ball. Mowing height accounts for 75 percent of the variation in the ball roll distance that arises from environment and course management factors in the summer. However, this influence is reduced to 50 percent in the spring.[2] Conversely, irrigation frequency is more important to the distance the ball rolls during the spring (24 percent) than it is during the summer (9 percent).[2] The proportion of nitrogen in the green also has an effect on the distance the ball will roll on the green during the summer but not during the spring. It has been speculated that low nitrogen rates may inhibit the leaf growth required to repair pitch marks, creating an uneven surface where the ball will bounce to a greater extent, reducing the distance the ball rolls.[2]

Summer

Mowing height 75%

Nitrogen 11%

Irrigation frequency 9%

Remaining factors 5%

Roll of the green

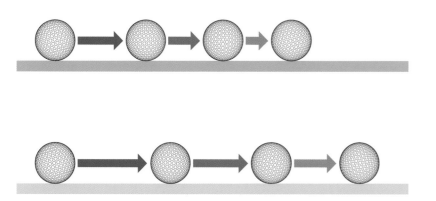

Types of grass The type of grass used on the green will affect the distance the ball rolls. From an analysis of five different types of grasses, the species that was shown to allow the furthest ball roll was Festuca rubra *ssp.* Litoralis. *The grass species that was shown to produce the least ball roll was* Poa annua. *This was the case for both wet and dry conditions.*[3]

How do altitude and latitude affect the golf ball?

Where could I go to hit the longest drive?

A golf ball's path is affected by aerodynamic and gravitational forces during its flight. The influence of both types of force is dependent on location. For instance, a golf ball will fly farther at higher altitudes because of a reduction in both aerodynamic and gravitational forces. The drag force will be smaller because the air density is lower at higher altitudes, and drag is proportional to air density (see Need to Know box on page 95) The lower air density also results in less lift force being generated but the latter's detrimental effect on driving distance is usually smaller than the gain from the reduced drag.

The decrease in gravity with increasing altitude might be more surprising. It is a common misconception that the force of gravity is constant on Earth. The gravitational force between two objects is influenced by the mass of the objects (in this instance the Earth and the golf ball) and the distance between them. The mass of the golf ball and the Earth are constant so the force of gravity will depend on the distance of the ball from the center of the Earth. By definition, gaining altitude takes you—and your ball—farther from the center of the Earth, so the influence of gravity will be lower.

The force of gravity also varies with latitude. The surface of the Earth is actually 13 miles (21 km) farther from the center of the Earth at the equator than at the North and South Poles. This difference arises from the spinning of the Earth around its central axis. The spinning creates forces that over time have changed the shape of the Earth from a sphere to an oblate spheroid, creating a "bulge" in the Earth's structure at the equator. Therefore, the ball will fly farther at the equator than at the two poles because the distance between the ball and the center of the Earth is greater—and the resulting gravitational force is smaller at the equator. So what are the implications of this for golf? If you want to increase your driving distance then you should play on a golf course that is at a high altitude and as close as possible to the equator.

Longest drive

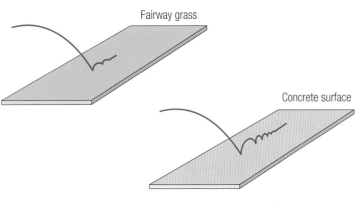

Fairway grass

Concrete surface

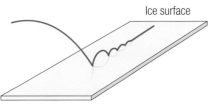

Ice surface

◄ **Effect of surface** *Officially, the longest ever drive on a golf course was struck 515 yd (471 m) in 1974. However, longer golf drives have been hit on airport runaways, where the surface is concrete, and even farther on ice. This is because the ball will travel farther once it has landed on concrete and ice than on the grass of a fairway. When the ball bounces it will lose less of its energy (due to a higher coefficient of restitution) on concrete and ice than on grass. After the ball has stopped bouncing, the reduced deformation of concrete means the rolling resistance will be reduced compared with that on grass— meaning the ball will roll farther on concrete. The ball travels so far on ice because the coefficient of friction is extremely low, even though the latter means the ball may slide more than it rolls.*

High and mighty[1]

Effect of altitude The altitude at which a golf drive is struck, relative to sea level, will have an effect on its distance. A drive struck at a high altitude will carry farther through the air than a drive that is struck at sea level. This is because of reduced gravity and also because the air density is lower, which reduces the drag on the ball during its flight and increases the distance it travels through the air. The lower air density at higher altitudes also reduces the lift force arising from backspin, and at very extreme heights this may result in a shorter carry—for certain combinations of launch parameters.

Launch angle 14.3°, ball speed 149 mph (240 km/h)

Carry (yd)	Altitude
263	14,335 ft*
262	10,000 ft
259	5,000 ft
253	1,000 ft
252	500 ft
251	Sea level

■ Carry distance (yd)

Launch angle 8°, ball speed 175 mph (282 km/h)

Carry (yd)	Altitude
294	14,335 ft*
303	10,000 ft
304	5,000 ft
297	1,000 ft
295	500 ft
293	Sea level

■ Carry distance (yd)

Shape of the Earth

Distance from the center of the Earth to the poles: 3950 miles (6357 km)

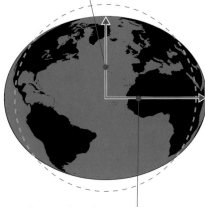

Distance from the center of the Earth to the equator: 3963 miles (6378 km)

Launch angle 11.5°, ball speed 167 mph (269 km/h)

Carry (yd)	Altitude
301	14,335 ft*
299	10,000 ft
289	5,000 ft
276	1,000 ft
274	500 ft
272	Sea level

■ Carry distance (yd)

▲ **Effect of gravity** The surface of the Earth is actually farther away from the center of the Earth at the equator than at the two poles. This has been caused by the planet's rotation, which has resulted in forces changing its shape from a sphere to an oblate spheroid. This extra distance from the center of the Earth to its surface reduces the effect of gravity on the ball—meaning the ball will fly farther in the air at the equator.

Launch angle 12°, ball speed 147 mph (237 km/h)

Carry (yd)	Altitude
237	14,335 ft*
239	10,000 ft
237	5,000 ft
233	1,000 ft
232	500 ft
231	Sea level

■ Carry distance (yd)

*14,335 ft (4369 m): altitude of La Paz Golf Club, Peru, the highest golf course above sea level in the world.

Golf coaches are often expected to dissect a golfer's swing technique with the naked eye and for a movement that is over within approximately 1.2 seconds. This is certainly a challenging task, even for a highly experienced coach. Therefore, the introduction of technology such as video cameras and performance analysis software has assisted the golf coach. The golf swing is one of the most difficult movements in sport, but recent advances in motion-capture technology, computer simulation models, and pressure measurement devices have allowed researchers to analyze its unique mechanics and offer invaluable insights into what a great swing "looks like." Making this wealth of data accessible to golfers and their coaches is the key to improving golfing performance. In this chapter, Robert Neal, a leader in the field of golf biomechanics, and Mark F. Smith apply their wealth of scientific knowledge to decipher the complexities of coaching technology.

chapter five

coaching with technology

Robert Neal and Mark F. Smith

Can 3D motion analysis improve golf performance?

How can 3D swing analysis improve my golf?

There are many benefits for the golfer who chooses to have their swing analyzed in three dimensions. Just think of the 3D analysis as a quantitative assessment of the swing. The "numbers" can be compared with an age-and-sex-comparable model and priorities for improvement can be readily determined. Tracking progress as the swing changes is a straightforward process using this type of technology. A further benefit of doing a 3D swing analysis is that physical limitations (such as stability, inter-segmental coordination and control, dynamic flexibility, strength, power, and so on) can be highlighted. This information can be used to underpin the prescription of exercises that are designed to change a golfer's

movements over time, improving the functioning body so that a more efficient and powerful golf swing can emerge.

To properly analyze the swing in 3D, the golfer must use a system that is capable of measuring all six degrees of freedom (DOF). This term is used by engineers to describe the six possible movement directions or axes in which a rigid body can move in 3D space. A simple way of thinking about this concept is to imagine the axes that the shoulder can move about: it can flex or extend, abduct or adduct (side to side movements), as well as undergo long axis rotation. These three axes of rotation would correspond to three DOF. In

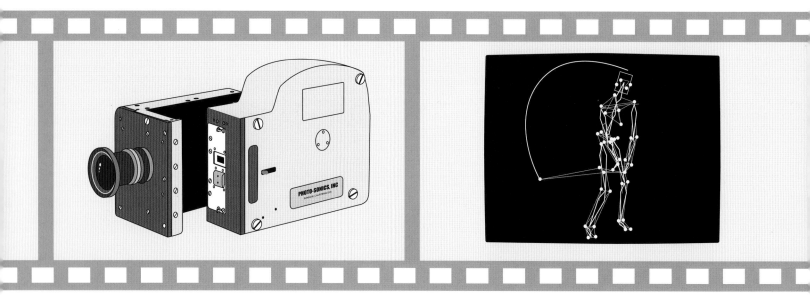

▲ *Candid camera* *In the 1960s and 1970s the Photosonics camera became the workhorse in biomechanics, since its features were perfect for scientific study. You could "phase-lock" multiple cameras, the film was pin-registered (this characteristic improved the accuracy and speed of digitizing later on) in the camera, lenses could be swapped easily, and you could achieve framing rates of 500 frames per second.*

▲ *Computer power* *In 1971, Abdel-Aziz and Karara[1] published their seminal work on the reconstruction of 3D position in space from 2D coordinates of the same points on film, and by this time computing power was sufficiently advanced to implement the mathematics. By the early 1980s, 3D studies were becoming more common as computing power continued to improve and researchers were prepared to increase the complexity of the models that were used to analyze human movement. Neal and Wilson[2] published the first paper on the 3D kinematics and kinetics of the golf swing.*

conjunction with these, the arm is also able to move sideways, thrust forward or backward and lift (up or down). These three translations represent the linear DOF.

The motion-tracking system depicted (below, right) uses the principles of electromagnetic induction to determine the position and orientation of the sensors attached to the golfer's body. The sampling rate for this type of hardware is 240 Hz, which means that the position and orientation of the body segments are gathered 240 times every second. Software is used to convert the sensor information to variables that are meaningful to the golfer and the golf coach. For example, the

amount of upper thorax (UT) or shoulder turn, in the golfer's language, can be measured to less than ±0.5 degrees with some systems.

Some 3D systems use optical methods to determine the position in 3D space of reflective markers (below, left). Most of these systems need at least eight or more—often 10 or 12—high-speed video cameras set up in a room in which the golfer hits shots. The golfer wears dark, tight-fitting clothing, and retro-reflective markers are placed on selected body points. These markers are tracked as they move in space and software is used to build an animation of the golfer as they hit shots.

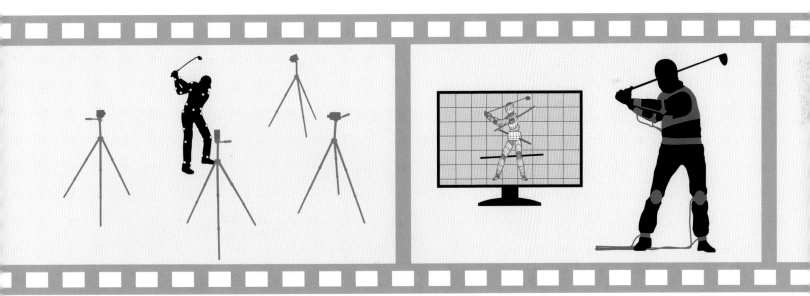

▲ **Video capture** *From the 1980s to today, multiple "video" cameras have been used along with retro-reflective markers to track the movement of these points on the body. Here you can see the cameras and the markers attached to a golfer as they hit shots (in practice, more than four cameras are used). The trajectories of these markers are then used to create a 3D "avatar." Sometimes other technologies are used (e.g. electromagnetic systems or rate gyroscopes, magnetometers, and accelerometers) to investigate golf swing mechanics.*

▲ **Super sensors** *Sensors or markers, depending on the technology being used, are attached to the golfer's body so that the movement (position, velocity, acceleration) of various body segments (e.g. hand, forearm, club, head, and so on) can be measured. Key positional information such as the address posture or impact position is provided for the golfer and these data can be compared with a range of acceptable values (given the sex, age, and experience of the golfer). Higher order kinematic data provides insight into the speeds and accelerations experienced by different parts of the body during the swing.*

Do ball flight data help the average golfer?

What can I learn from watching my ball in flight?

Ball flight information is extremely valuable to the average golfer because the player can, through observation, deduce what the club was doing at impact. The trajectory of the ball, the sense of the impact location of the clubhead, and the shape of the divot (when appropriate) taken together can give you most of the clues you need to determine, with a high degree of certainty, the way the clubhead moved to cause the ball to fly that way.

First, the golfer must have an understanding of some basic principles of impact, spin, and trajectory of a ball so they can take full advantage of any information they gather. Impact between the ball and club lasts for a tiny fraction of a second—less than 1/2000th of a second to be precise—yet during this time the club transfers all the necessary information to send the golf ball on its way. The impulse applied by the club, coupled with the forces of gravity and air resistance, determines the trajectory of the ball during flight.

By applying the basic rules governing ball flight, a player can begin to deduce reasonably accurately what the motion of the club was at impact. For example, a player watching their ball fly in a straight trajectory (with no sideways curve), but to the left or right of the intended target, can conclude that the clubface angle and the club path were aligned—or collinear—but not correctly oriented with respect to the target. Similarly, if the initial direction of the

ball is toward the intended target, the player can conclude that the face orientation at impact was correct. Understanding ball flight also helps when hitting from the rough—the ball will not curve in a slice or hook as much. In other words, most golf shots fly pretty straight when hit from the rough. Lastly, short irons produce more backspin on the ball than a driver and, as a consequence, the ball will not curve as much in flight if the club path and face angle are not collinear.

Intended and actual ball flight

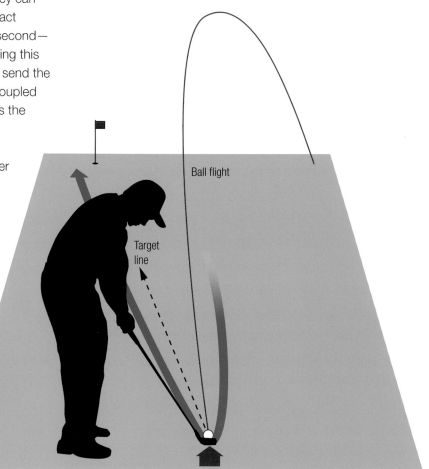

Ball flight

Target line

▶ **Slice** *The red arrow represents the direction in which the clubface was pointing at impact and the blue arrow is the club path. Because the path is more to the left (out-to-in) than the clubface angle, the ball curves to the right (a slice for this golfer). An impact near the heel of the club will exacerbate the slice spin applied to the ball.*

*▼▶ **Loft*** *Viewing the impact between the clubhead and ball side-on, the most important physical parameters to understand are the club delivery data (v_c and α) along with the ball launch conditions (v_b, β, and ω_b). When there is a difference between the angle of attack (α) and the dynamic loft of the club (θ), then spin (ω_b)—probably backspin—is created on the ball.*

Ball loft

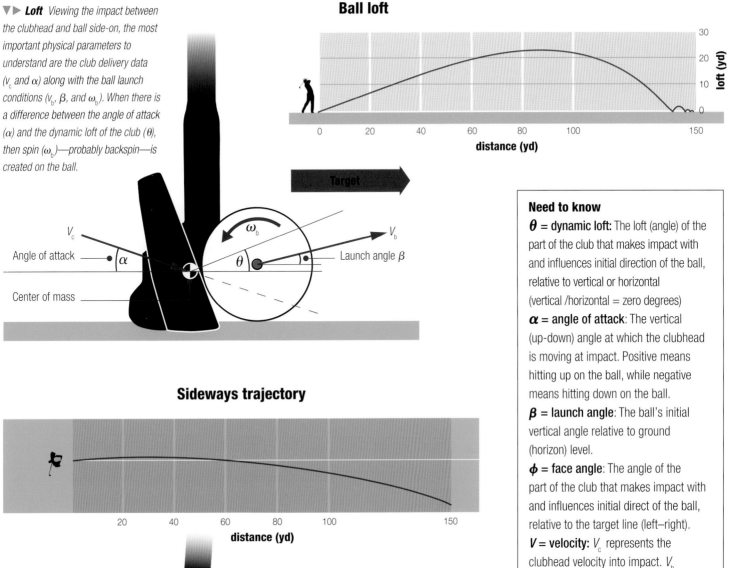

Angle of attack — α

Center of mass

V_c

ω_b

θ

Launch angle β

V_b

Target

Sideways trajectory

distance (yd)

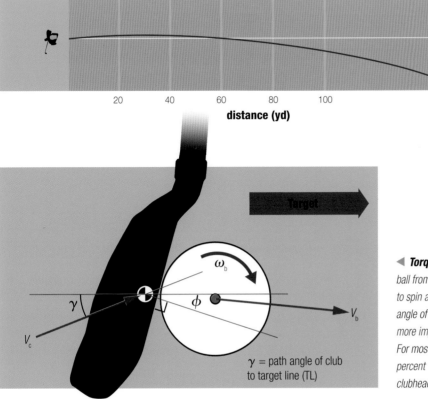

Target

γ

V_c

ω_b

ϕ

V_b

γ = path angle of club to target line (TL)

Need to know

θ = dynamic loft: The loft (angle) of the part of the club that makes impact with and influences initial direction of the ball, relative to vertical or horizontal (vertical /horizontal = zero degrees)

α = angle of attack: The vertical (up-down) angle at which the clubhead is moving at impact. Positive means hitting up on the ball, while negative means hitting down on the ball.

β = launch angle: The ball's initial vertical angle relative to ground (horizon) level.

ϕ = face angle: The angle of the part of the club that makes impact with and influences initial direct of the ball, relative to the target line (left–right).

V = velocity: V_c represents the clubhead velocity into impact. V_b represents ball's initial velocity.

ω = spin rate: ω_b represents the angular velocity around the ball's center.

*◀ **Torque*** *Similarly, when viewing the collision of the club and ball from above, a torque will be applied to the ball that causes it to spin and eventually deviate sideways. Of the two factors, path angle of the club (γ) and face angle (ϕ), the face angle is much more important in determining the starting direction of the ball. For most impacts, including putting, face angle accounts for 85 percent of the ball's initial direction of movement and the clubhead path accounts for the remaining 15 percent.*

121

How can golf apps improve swing dynamics?

Can I use my smartphone to improve my golf?

Over the last 20 years, video technology has grown in both popularity and sophistication. Increasingly being used by top coaches to study the swings of the best players in the world, video has helped coaches understand what good players do (and don't do) and so has helped inform their teaching. The advent of good-quality cameras in smartphones and golf-swing software in the form of downloadable apps offers the golfer a compact video–capture–and–analysis package that could, if used appropriately, provide some benefit in helping to improve their game. But is there evidence that supports the effectiveness of these apps?

Interestingly, there is mixed experimental evidence as to the real benefit of video analysis and feedback in improving motor performance for movements such as the golf swing. Certainly, findings taken from other sports, such as tennis[1–2] and swimming,[3] and from other motor skills such as surgery,[4] seem to suggest that video feedback has no meaningful impact on skill acquisition (the ability to learn and then perform a task consistently and accurately) compared with traditional verbal or instructional feedback. On the other hand, research has also revealed that gymnasts[5] and golfers[6–7] have benefited from comparing video footage of themselves to footage of more skilled performers. Evidence also suggests that when coaches watch video footage they are more likely to empathize with their player's thoughts and feelings, and therefore the use of this technique can facilitate a more supportive relationship between teacher and player.[8]

Taken together, there does appear to be some benefit of video capture in improving performance, but it is dependent on the extent of understanding and practical application a player possesses. For a player wishing to make best use of this easy-to-access technology there are some fundamentals that must be followed. Dr. Robert Neal, a world-leading golf biomechanics specialist and authority on swing analysis, insists that for video footage to be useful the player must have sufficient knowledge of golf swing mechanics to undertake some basic analysis of their technique by themselves. In other words, the player must know what they are looking for and must be consistent in their approach to filming. It probably makes sense to have the objective eyes of a teaching professional do the analysis initially, but once the player understands what needs to be monitored, the smartphone video can prove to be extremely valuable.

An app for that

▲ *Get smartphone Research suggests that using video feedback to gather information about a player's swing can improve the player's effectiveness.[7] When using a smartphone app for this purpose, it is important to ensure that the player is filmed both down the target line (TL) and face-on (FO). Also, if looking at the video without a coach, it must be clear to the player what they are looking for—for instance, head position or spine angle—and what they wish to improve. Thus, they need to know how to apply the changes correctly.*

Pocket coach

Highly skilled player at address

Highly skilled player at impact

Highly skilled player post-impact

◀▲ ***Phone for help*** *Some of the key differences between the way highly skilled and low-skilled players swing the club and move their bodies can normally be picked up when using video capture technology from a smartphone. The image shows a highly skilled player captured at different times during the swing—the red lines highlight face-on/target-line positioning as shown on the app.*

▼ ***Key differences*** *The application of decades' worth of evidence, on-course observation, and technological advancement has enabled detailed differences between skill levels to be quantified. The most common differences are summarized in the table below.*

Variable	Highly skilled player	Low-skilled player
Pelvis sway	No sway away from the target on the backswing. Center of pelvis moves approximately 4–5 in (10–13 cm) toward the target on the downswing.	Pelvis sways away from the target on the backswing (often of the order of 2–6 in/5–15 cm). It does not sway laterally toward the target on the downswing.
Club path/ swing direction	The trajectory of the center of the clubface on the downswing from the time that the shaft is parallel to the ground "matches" the trajectory on the follow-through, as viewed from down the target line (neutral club path).	The trajectory of the center of the clubface on the downswing from the time that the shaft is parallel to the ground is "above" the trajectory on the follow-through, as viewed from down the target line (out-to-in club path).
Spine angle	Maintained through the backswing, downswing, and early follow-through. When viewed from down the line, the pelvis does not thrust toward the ball.	Spine angle changes as the pelvis thrusts forward, toward the ball. This phenomenon commonly occurs on the backswing and then increases on the downswing.
Head sway and lift	Head moves a few inches away from the target and usually drops a little on the backswing. On the downswing it drops a little farther and moves laterally toward the target so that it is close to where it was at setup.	Head either sways excessively or remains too still as it lifts up on the backswing. On the downswing, it often rises more and is commonly closer to the target (in the sway direction) than it was at setup "in front of the ball" in golf parlance).

SCIENCE IN ACTION pacing the effort

It is easy to forget that golf is not just about hitting accurate shots consistently. A coach can also help with the physical and mental strategy involved in playing a whole round or, for top players, a tournament. Pacing is playing at a steady rate in order to avoid overexertion or underperformance.

Literally meaning a rate of movement or progress, and traditionally used as a unit of measurement, "pace" originates with the ancient Roman soldiers who marched in paces to ensure they kept their ranks with precision.[1] This practice continued in most European infantry armies for centuries and, given the physical and mental importance of entering battle in an optimal physical condition, the regulation of pace during such marching would have been pivotal in ensuring overexertion was minimized and essential metabolic reserve preserved.[2] Within the context of a round of golf, a golfer's physical pacing must ensure they are able to maintain their effort around the course, staying fresh for each shot.

More importantly, however, is a golfer's use of mental energy. Seeing the course as 18 distinct sections ensures they have a planned strategy for each hole: having decided where to place the ball, making the shots they feel comfortable playing, knowing when to take a risk, finding ways to feel relaxed throughout the round, and not dwelling on past shots. This conscious act of planning a set of actions designed to achieve the goal of a low score—the strategy—requires knowledge of the course so that a player can complete the round on a shot-by-shot basis without unnecessary physical effort and mental strain.[3,4] Using a course planner to set out your strategy, knowing your distances, regularly referring back to these in anticipation of your next shot, and managing your physical and mental reserves will all help you to get to the 18th green in the same shape in which you left the first tee.

▶ **Water works** *South Korean Tour pro K. J. Choi takes on fluids during the PLAYERS Championship at TPC Sawgrass. Maintaining appropriate fluid and nutrition levels out on the course helps to ensure that a player stays in peak physical condition throughout a round. It has been proven that doing this also contributes to sustaining mental focus and alertness, essential to good decision-making during play, for individual shots and planning your whole round.*

How can movement quality during the swing be assessed?

What methods are available to help me analyze my swing?

In golf, there are three main technological methods for providing three-dimensional representations of golfers' shots: optical systems, electromagnetic solutions, and microelectricalmechanical systems (MEMS). Optical and electromagnetic systems are true 3D measurement technologies, while MEMS are "pseudo-3D." True 3D systems are capable of measuring the full six degrees of freedom (6DOF) required for the description of movement of rigid bodies in 3D space. The pseudo-3D systems only measure three degrees of freedom, usually giving information on the orientations of body parts, but not their positions in space. The key benefit of the 6DOF systems is that these systems give the complete picture without the necessity of running other technologies. Optical systems track reflective markers on selected landmarks of the body by automatically distinguishing the markers on the video image from the background. The horizontal and vertical coordinates of the markers as "seen" by multiple cameras are used to compute their 3D trajectories. A calibration procedure is necessary prior to data capture to obtain internal camera parameters for completing the transformation from 2D to 3D. Additional software is used to compute the 3D rigid-body motion, using previously defined relationships among the markers. Electromagnetic systems rely on the strength of induced currents to determine the position and orientation of sensors attached to the golfer's body relative to the source of varying magnetic fields. MEMS uses a combination of rate gyroscopes, accelerometers, and magnetometers (which determine the angular position of the Earth's magnetic field) to accurately measure the orientation of sensors attached to the player's body. These systems can provide precise information on the rotational degrees of freedom but give no information on the position of the sensors in space. Systems based on these technologies require additional input for a coach or player to make comprehensive assessments of movement quality.

Movement sensors

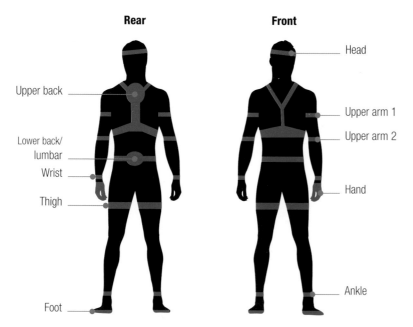

Rear

- Upper back
- Lower back/lumbar
- Wrist
- Thigh
- Foot

Front

- Head
- Upper arm 1
- Upper arm 2
- Hand
- Ankle

◀ *Electromagnetic system* *Today's state-of-the-art electromagnetic systems offer a complete 16-body-segment model (using 12 sensors). Recent advances in the technology mean that the system can be almost wireless, with the sensors connected to a smartphone-sized device that clips to the belt and transmits the data via a radio link to the computer. Sensor currents are read and processed by the electronics unit before the data are sent to the computer. Complete signal processing and transmission occurs in less than 10 ms. The strengths of these systems are the wireless transmission of data, high sampling rate (from 120 Hz wireless to 240 Hz if tethered), simplicity of calibration, feasibility of attaching sensors to the club, possibility of full body-rendered 3D animation in real time, and that the systems are portable, quick to set up, and can be used indoors or outdoors. Disadvantages are that movement must be within the calibrated space of the transmitter (a volume of approximately 71 cu ft (2 m^3) to ensure accuracy of measurement) and that large metal objects can interfere with data from the unit.*

Big screen simulation

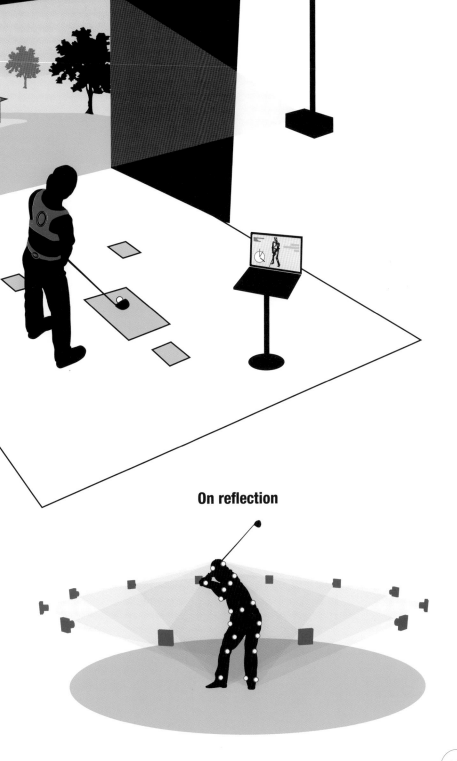

▶ **Indoor setup** Wireless
sensors are attached to the body
segments and software is used to
convert the electrical signals into
orientation data. These data are transmitted
via radio link to the computer for additional
processing and conversion into graphic
representation—In this case indoors, projected
onto a screen in front of the golfer, offering
real-time feedback.

▶ **Optical systems** These use reflective markers placed
on the golfer to reconstruct 3D movement. Once the marker
trajectories are known, software can create an animation of
a rigid-body representation of the body and calculate various
kinematic parameters. Key advantages of this system are that
there are no cables, high sampling rates are possible (up to
500 Hz), calibration is anatomically referenced, and full
body-rendered 3D animation is possible. The disadvantages
are that most systems must be used indoors, with subdued
lighting, the calibration process is complicated and takes
a long time, and no real-time visual feedback is available.

On reflection

127

equipment: the wedge

Who isn't amazed at the skill of a Tour player who can hit the ball from 75 yards (68.5 m) off the green and spin it back 10 ft (3 m) to a tight pin position, or who can gently flop the ball over a trap with just a stride's length of green to play with? The best players in the world are able to demonstrate such control and precision with their short game because, first and foremost, their swing technique is impeccable. However, a good player should also recognize that their weapon of choice can make a difference when it comes to being inch-perfect around the green.

Prior to the 1930s, a single pitching wedge—commonly known as a "jigger"—emerged to assist the golfer in recovering from tricky positions, such as the sand. With reduced loft to prevent digging into soft ground, the low launch angle had the unintended consequence of creating high resistance when attempting to dig the ball out from buried lies. In 1931, Gene Sarazen evolved the wedge design, thereby creating the "second wedge," or sand wedge. For several decades following its initial invention, the standard pitching wedge maintained a loft of 48–50 degrees while that of the sand wedge was slightly higher at 54–56 degrees. It wasn't until the 1980s that proponents of the short game seriously began to rethink the wedge design with even greater loft. Innovators, with scientific minds, realized that the traditional wedge design wasn't adequate to handle the increasing number of new courses featuring elevated and undulating greens. The development of higher-loft clubs with reduced bounce enabled access to a larger number of tight pin positions.

While the driver, iron, and putter have benefited from significant leaps in technological development, the wedge has lagged behind slightly, receiving the lowest amount of engineering ingenuity. It is unlikely that this has been missed off the major manufacturers' to-do list, so it is more probable that the wedge is pretty much fit for purpose as it stands. However, scientifically informed innovation has brought some important changes to the wedge to help the player, including variations in head design for a lower center of gravity, improved sole shape to reduce bounce, and features to increase ball control and spin rate off the face.

Loft conversion

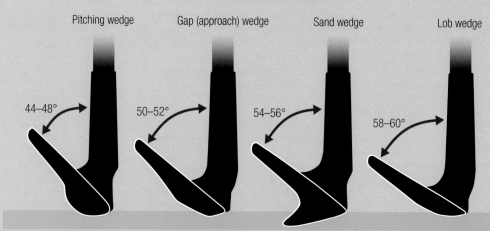

Pitching wedge 44–48° Gap (approach) wedge 50–52° Sand wedge 54–56° Lob wedge 58–60°

◀ **Typical lofts** *Modern wedges are available in just about every loft from 46–64 degrees. Although generally wedge lofts come in even numbers, each manufacturer has a different take on what the standard loft is for each of the four wedges.*

Spin doctor

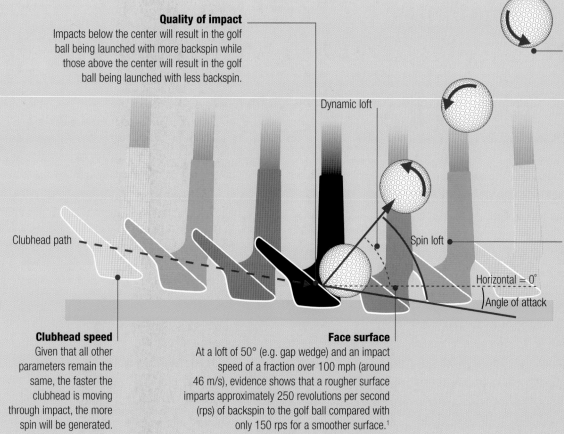

Quality of impact
Impacts below the center will result in the golf ball being launched with more backspin while those above the center will result in the golf ball being launched with less backspin.

Dynamic loft

Ball
A softer covered three-piece ball results in creating a greater coefficient of friction. At an impact speed of 83 mph (37 m/s) the measured backspin of a three-piece ball has been recorded at 178 rps for an impact angle of 45° (e.g. pitching wedge). The corresponding value for a two-piece ball was 171 rps for the same impact angle.[1]

Clubhead path

Spin loft

Horizontal = 0°

Angle of attack

Spin loft
For a given clubhead speed, changing the relationship between the dynamic loft and attack angle in theory alters the spin loft and therefore the spin rate of the ball. Although moving the ball back in the stance generally creates a more negative attack angle, the dynamic loft will be offset by a similar amount, resulting in an unchanged spin loft and therefore unchanged spin rate.

Clubhead speed
Given that all other parameters remain the same, the faster the clubhead is moving through impact, the more spin will be generated.

Face surface
At a loft of 50° (e.g. gap wedge) and an impact speed of a fraction over 100 mph (around 46 m/s), evidence shows that a rougher surface imparts approximately 250 revolutions per second (rps) of backspin to the golf ball compared with only 150 rps for a smoother surface.[1]

▲ **Spin** *The amount of spin imparted to the golf ball by the wedge is affected by a number of factors—the face surface of the wedge, the path of the clubhead as it approaches the ball, the loft of the wedge, the speed of the clubhead at impact, the specific quality of impact, and the ball itself.*

▼ **Bounce** *The bounce of a club is defined as the angle, in degrees, between the ground and the club's sole plane. At impact in soft turf or sand, a higher bounce angle increases the upward resisting force on the sole of the club to prevent excess digging. For firm turf or sand, lower bounce angles improve contact by reducing this resisting force, allowing the leading edge to easily slide underneath the ball.*

Club bounce

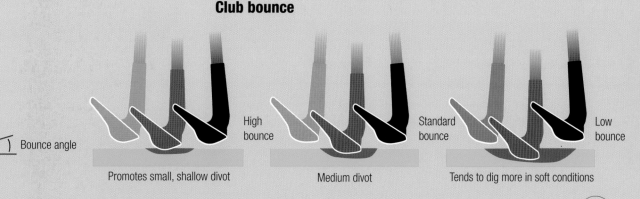

Bounce angle

High bounce

Standard bounce

Low bounce

Promotes small, shallow divot

Medium divot

Tends to dig more in soft conditions

What are the advantages of coaching technologies to the coach?

Can a coach with the latest technological tools improve my golf?

Going back not that many years, golf coaches had to rely only on their own sensory inputs—watching and listening to a golfer—and then filtered the necessary information from the unnecessary. Today's golf coaches have access to a multitude of different technologies which can provide information to guide their coaching decisions. So are today's coaches any better because of the technological devices now available to them? It can be argued that having lots of different types of data from various technologies does not make you a good coach. In fact, bombarding a player with unimportant information might be detrimental to helping them learn how to improve their golf. The golf coach must sift through the available information for the most important feedback. It is true, however, that putting high-tech devices in the hands of a good coach will allow them to make more rapid progress with a player. Furthermore, the feedback from high-tech devices, if provided in a systematic way, can help the golfer to understand their swing better and gain a feel for a correct pattern of movement more rapidly than they would be able to without access to such data.

There are many new tools which help a player and their coach to focus on improving technique. Very high-speed cameras and "launch monitors" now enable the recording of the impact of the club and ball—an event which takes just 450 ms. Tracking systems are being used to understand the movement of a ball after impact. Using radar, the trajectory, speed, and even level of spin of a ball can be mapped, providing a player with detailed feedback on the outcome of their swing. Putting technique can also benefit from tracking systems—they allow the coach to measure the motion of the putter during the stroke, particularly as it approaches the ball. The coach can monitor the angle of the putter head at impact to determine if the player is "cutting," "hooking," or squarely hitting the ball. Obviously, armed with such knowledge, the good coach can help the player make corrections to technique in order to improve the quality of impact.

On track

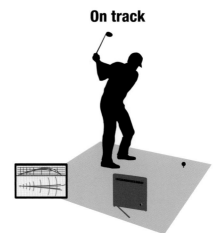

◀ *Tracking devices* According to ball flight laws, the club path and face angle are key parameters in determining the trajectory of a shot, and it is possible to analyze them using modern imaging equipment. The latest devices enable a golfer to monitor the swing, clubface angle, and ball launch to the ball's trajectory, all in real time.

Putting technique

◀ *Sonar* When evaluating putting technique, it is impossible for the golfer or coach to know the face angle and path of the putter when the putter makes contact with the ball. Sonar or electromagnetic tracking systems allow the coach to measure the motion of the putter during the stroke and as it approaches the ball.

Swing forces

◀ *Force plate* The force plate differs from the pressure mat in that it typically measures the three components of the ground reaction force, or GRF (medio-lateral, antero-posterior, and vertical), the center of pressure, and the free torque applied to the ground. This helps the golfer to understand weight transfer and shear forces during the swing.

High-tech feedback

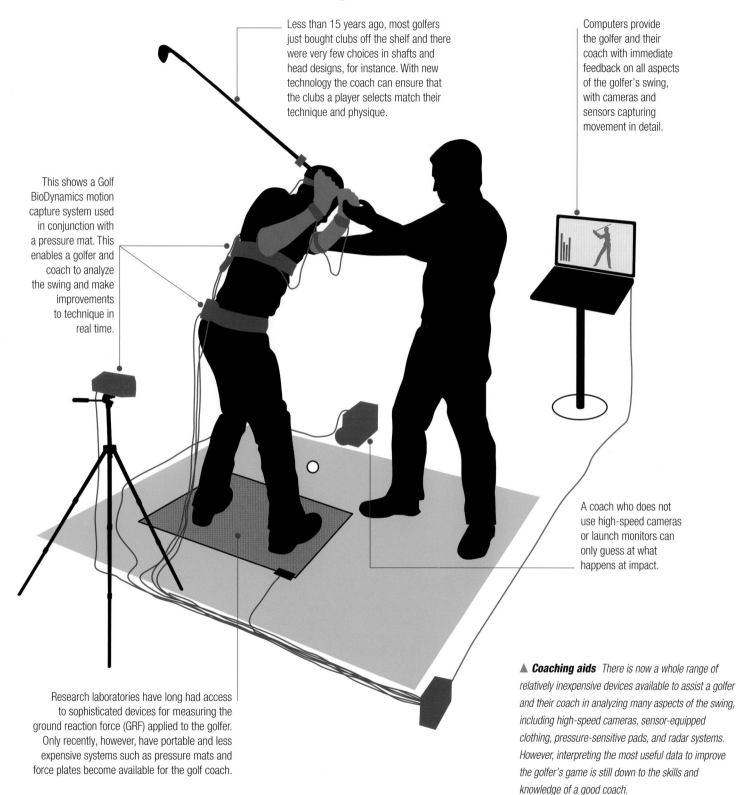

Less than 15 years ago, most golfers just bought clubs off the shelf and there were very few choices in shafts and head designs, for instance. With new technology the coach can ensure that the clubs a player selects match their technique and physique.

Computers provide the golfer and their coach with immediate feedback on all aspects of the golfer's swing, with cameras and sensors capturing movement in detail.

This shows a Golf BioDynamics motion capture system used in conjunction with a pressure mat. This enables a golfer and coach to analyze the swing and make improvements to technique in real time.

A coach who does not use high-speed cameras or launch monitors can only guess at what happens at impact.

Research laboratories have long had access to sophisticated devices for measuring the ground reaction force (GRF) applied to the golfer. Only recently, however, have portable and less expensive systems such as pressure mats and force plates become available for the golf coach.

▲ **Coaching aids** *There is now a whole range of relatively inexpensive devices available to assist a golfer and their coach in analyzing many aspects of the swing, including high-speed cameras, sensor-equipped clothing, pressure-sensitive pads, and radar systems. However, interpreting the most useful data to improve the golfer's game is still down to the skills and knowledge of a good coach.*

Can biofeedback advance golf performance?

What is the future of golf coaching technology?

So what does the future hold for golf coaching technology? And how may recent scientific discoveries begin to shape the way the golfer trains and competes? The possibility of a fully integrated system of sensors and feedback devices, positioned over the body of the player, which allows both coach and golfer to access real-time biofeedback in the field, is not far-fetched. Biofeedback, in a practical context, is defined as a training program in which the golfer is given information about movement (for instance, position or orientation of a body segment, speed of movement, and coordination) which is not normally available, with the goal of improving swing kinematics. The feedback can be auditory or visual, the benefit of auditory real-time feedback being that the golfer can monitor the feedback and still watch the golf ball. The most important and underlying objective of any form of biofeedback provided to the golfer is that they can gain the correct feel (in their body) of the desired movement, without any external forces being applied by an instructor. In golf coach-speak, the player can quickly learn to "own" the new movement pattern.

From a scientific perspective, biofeedback is the process of gaining greater awareness of the many psychophysiological systems through the accompanying responses that occur during movement. Primarily using instruments that provide information on the activity of those systems, the objective is to increase voluntary control over physiological processes that are otherwise outside awareness, using information about them in the form of an external signal. Some of the processes that can be monitored during golf include brainwaves, muscle tone, skin conductance, heart rate, pressure patterns, and pain perception.[1-7] Taken together, biofeedback may be used to improve performance and control the physiological changes which often occur in conjunction with changes to thoughts, emotions, and behavior. Eventually, the golfer may be able to maintain such changes without the use of extra equipment, so that they become automatic.

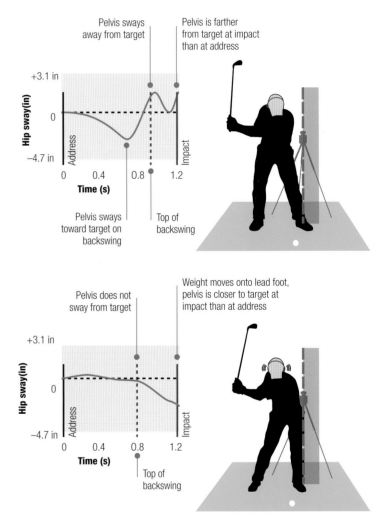

Pelvis movement

▲ **Audio aid** *Auditory biofeedback can be used to change a golfer's movement pattern. At halfway through the downswing (top) the pelvis has rotated too much and not translated sufficiently toward the target. The golfer would know this immediately because no beep would be heard. Below, the golfer would have heard the beep, letting him know that his hips have moved far enough toward the target—into the indicated zone—and that he can begin rotating his hips as rapidly as possible. Equally, if he had moved his hips too far to his left the beep would have gone off, which is a disadvantage of the system.*

Head movement

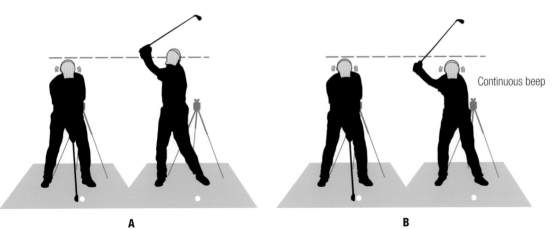

A   B

Continuous beep

◀ **Head down** The golfer initially lifted his head up too much on the backswing in A. By using a "virtual ceiling" placed over his head, the player can learn to change his movement. Starting his backswing with the audio tone on, the player has to keep it on during the entire backswing in B. He quickly learns what it feels like to maintain the correct positions, including spine angle.

Future training aids

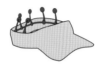

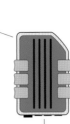

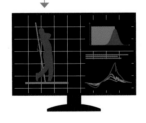

◀ **Biofeedback future** So what may the future of golf biofeedback technology hold? Although yet to be fully realized, emerging research is leading to the evolution of complete biofeedback systems that assist in the optimization of golf performance.[1–7]

The visor acts as a multi-system device, monitoring, recording, and then feeding back the most vital data to the golfer.

Micro-sensors and gyroscopic devices detect body position, muscle activation patterns, breathing response, heart rate, and skin temperature.

Sensors within the clubhead and shaft provide instant feedback on swing speed and impact position.

An "intelligent training pod," worn on the belt, collates data measured by on-body wireless devices. Feedback in the form of coaching advice, vibrations, or sounds can be delivered through the visor.

Each smart shoe uses ground contact sensors to measure pressure and force information throughout the swing.

A fully responsive online software package collates all golfer data to build up an individual training program based on the player's strengths and weaknesses.

To become a better golfer requires time, effort, and, of course, plenty of patience. With it taking roughly 10,000 hours of purposeful practice to hit the lofty heights of real golfing success, the search for ways in which practice can become more effective, time-efficient, and individualized has led to science for the answers. A good golf practice strategy should be open to change and be continuously adjusted as skills, time, and equipment vary. As the player's game changes, so will strengths and weaknesses. Practicing the golf game properly is vital to improving. Knowing how to practice increases self-confidence, reduces scores, and lowers golf handicaps. It also increases the fun of playing as scores drop on each round. John Hellström and Mark F. Smith consider how science has aided our effective implementation of practice regimes to further golf performance.

chapter six

the practice process

John Hellström and Mark F. Smith

Does "deliberate practice" improve golf scores?

Can I improve my golf just by playing more often?

When watching the world's best players compete for top places at the Majors, the extensive training each has undertaken doesn't necessarily spring to mind. Such expertise develops over an extended period of time.[1] A training period of 10 years or some 10,000 hours may be needed to become an expert, as first suggested by studies on chess players and musicians.[2,3] Similar amounts of training are also needed to reach an international level in some sports.[4] Neurophysiological evidence indicates that the brain slowly transforms as a consequence of practice.[5] Each time a pattern of activity is repeated, new connections between brain cells, called synapses, are created, reinforcing neural pathways and enhancing the player's ability to perform that activity automatically.

Given that the average golf handicap index is 16.1 for men and 28.9 for women, according to the United States Golf Association, many are far away from being competitive at high level. Most players have stopped lowering their average scores well before they could shoot under par regularly. Research has started to indicate that piling in the hours on the course or practice green alone may not be enough to improve. Even at a young age, playing for fun and playing to win need to be carefully managed to maximize the golfer's chance of long-term success—having fun is important for the juniors beginning a new sport in order to maximize participation and minimize dropout.[6,7] Early specialization may not be needed in golf, where top performers often are in their mid-30s.[5]

▶ *Structured learning* *As a beginner starts to play golf, opportunities to engage in unstructured free play and practice often take up the large majority of available golf time. This is termed the sampling phase and is not governed by detailed score monitoring, performance analysis, or too much critical review of performance—in essence, the main goal is having fun.[9,10] The gratification is immediate and there is no focus on correction. If a player wishes to advance and commit, the number of activities decreases and the type of practice begins to change. The volume of play typically decreases and deliberate practice increases. This means that golf gradually becomes the top priority, with clear goals, carefully monitored training, and focus on feedback and immediate correction (known as the investment phase).*

▪ Play
▪ Deliberate practice

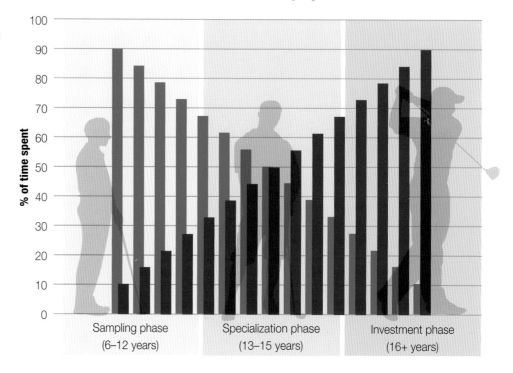

Growth of a player

% of time spent

Sampling phase (6–12 years) · Specialization phase (13–15 years) · Investment phase (16+ years)

What is clear is that those golfers who avoid becoming stale and unimaginative in playing and practicing are able to continually improve by constantly challenging themselves. They set relevant goals, often with the assistance of a dedicated and supportive coach, and create training situations in order to help them exceed their current skills, encourage high focus during training, and enable evaluation of their performance.[8] This is called deliberate training,[3] and players need a high volume of this as they approach adulthood if they are to reach an international level in golf. Choosing the right training system, having the support, the right mindset, and plenty of time to practice are the most effective ways to lower that handicap.

▼ *Warm-up* *A critical challenge for all golfers wishing to improve is to avoid "arrested development." The concept, coined by Anders Ericsson, the father of "deliberate practice," refers to a period of player development associated with automaticity and staleness instead of continued learning and improvement. Actively seeking out demanding on- and off-course challenges during practice increases the performance level and lowers scores.*

■ Golfers who become world class players continue to improve

■ Amateur players who stop improving (arrested development)

Deliberate practice makes perfect

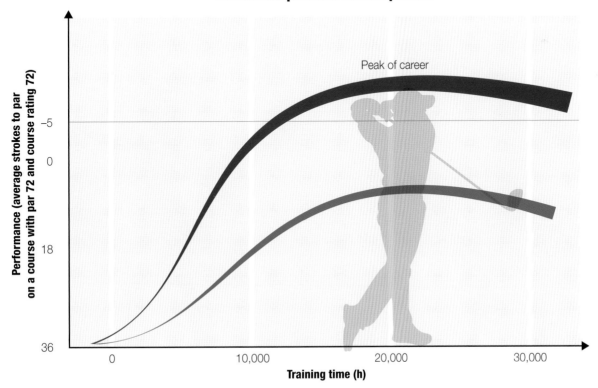

What are the psycho-physiological effects of a warm-up routine?

How does a warm-up before practice help?

Successful performance is built around routine—every great golfer has one. From the pre-shot ritual to the post-game analysis, building in repetitive patterns of positive behavior aids a golfer's preparation, on-course performance, and post-match reflection. A common routine within the professional golfer's armory is the essential, but often misunderstood, warm-up. Although not often shown in TV coverage, it's easy to spend many hours at tournaments observing the top players on the practice ground. They work through the array of shots, calibrating their body and mind, and deciding final swing thoughts with their coach. This important "before-play" aspect works on a number of mental and physical levels and all players can benefit.

As is often mentioned in popular coaching books and player manuals, one purpose of a warm-up is to improve the dynamic function of muscles, by elevating localized blood flow, increasing temperature, and activating key enzymes. Collectively, these processes get the muscles in a state of "readiness." There is also evidence to suggest that undertaking regular golf-specific stretching before play can reduce the incidence of muscle strains.[1,2]

Research indicates that a warm-up employing dynamic stretching—an active movement that pulls the muscle into an extended position without exceeding its ability—results in greater clubhead and ball speeds than if the stretches are done

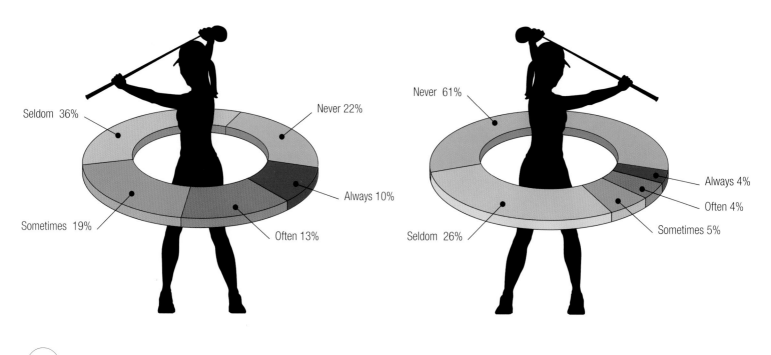

Percentage levels of warm-up exercise by golfers preceding play

Seldom 36%
Never 22%
Always 10%
Sometimes 19%
Often 13%

Percentage levels of warm-up exercise by golfers preceding practice

Never 61%
Always 4%
Often 4%
Sometimes 5%
Seldom 26%

Benefits of a warm-up

without movement or not at all. Additionally, this research found that players who undertook regular, golf-specific dynamic stretching before play achieved a higher number of centralized impacts on the clubface than the static stretch group.[9] Taken together, this equates to greater distance and more consistency.

Devoting a little time before each round to a brief routine can also have a beneficial effect on the way the mind prepares for play. Studies of highly skilled tennis players have indicated that a warm-up consisting of practice swings can optimize arousal level and activate a player's ability to engage in positive mental imagery.[4] Given that mental imagery can be an effective way of reducing performance-related anxiety, increasing concentration and focus, and removing unwanted thoughts, the warm-up process may serve as a way to calibrate the mind as well as prepare the body.

◀ *Practice makes perfect* *Whether preceding practice or play, evidence from studies on amateur golfers with an average skill level reveals the likelihood of a golfer performing a warm-up routine is generally low.[3, 6] Given the advantages to both mind and muscle, even the briefest activity—built regularly into a player's time at the course—could have significant benefits for their performance.[9]*

All in the mind?

1 Research indicates that a warm-up routine of practice swings and light exercises puts the golfer's mind in the right place and can lead to better performance.

2 Increases in blood flow to the brain enhance the nervous system, activating alertness and cognitive function.[4]

3 Additionally, rehearsal of the desired swing movements at an optimal velocity may act to facilitate coordinated motor pathways and improve tempo and rhythm.[5]

Or more in the muscle?

1 A general body warm-up can increase muscle blood flow and body temperature, which speeds up contraction of the important fast fibers activated during the golf swing.[7]

2 Static and dynamic stretching has been shown to alter the biomechanical length–tension relationship of shortened, tight muscles.[8]

3 Dynamic stretching has been shown to increase flexibility of the muscle–tendon units, helping the golfer to obtain the desired biomechanical dynamics throughout the swing.[10]

How does a golfer learn a new swing?

→ How soon can I play without thinking about my swing?

Golfers who are starting their playing journey are often in awe of how easy it looks when skilled players swing. But how do they get to such a stage of performance? Cognitive psychologists have pondered such questions for over a century and have suggested a number of theories. One, which has gained popular support within sport science over the last 50 years, is that proposed by the renowned Fitts and Posner—a phasic conceptualization of motor skill learning evolves in the learner through three stages: the cognitive phase (understanding the swing); the associative phase (refining the swing); and the autonomous phase (making the swing automatic).[1]

In the early stages, there are typically many errors in the movement, with swing characteristics and ball flight being highly variable. The player searches for answers based on the basic fundamentals of the game and movement.[1] Learning the golf swing demands a lot of attention and concentration, and players do not generally know what to correct in order to improve. They therefore need specific instruction, demonstrations, and other verbal information that will support their development. Research has indicated that those who learn to play golf at an early age may learn more implicitly than those learning at later ages.[2] Young juniors may rely on

Process of deliberate practice

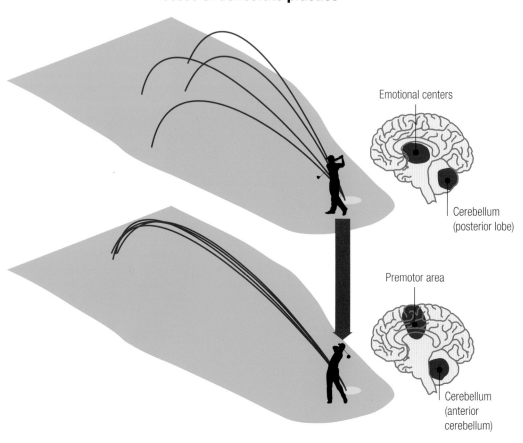

Emotional centers

Cerebellum (posterior lobe)

Premotor area

Cerebellum (anterior cerebellum)

▶ ***Flow of learning*** *Coaches and players all face the flow of learning whenever they come to develop a golf swing. Practice, whether coached, instructed, or self-initiated, leads players through a number of continuous periods of learning to true behavior change.[4] The extent of change, however, depends on a number of interconnected factors such as an individual player's training history, biological age, instructional quality, and the nature of practice.*

◀ ***Brain training*** *Deliberate golf practice induces changes in the gray matter of the brain,[5] with changes being dependent on the learning stage the player is moving through.[6] The posterior lobe of the cerebellum and emotional centers have a higher involvement during the early stage of motor skill learning,[7] which may reflect the evolution of a rhythm for the movement sequence and conscious "emotional" thought.[8] When the rhythm has formed and become autonomous, the anterior cerebellum and premotor area become more active.*

Learning progression

Stage one

Stage two

Stage three

Stage four

Unconscious incompetence
This is where any learner starts—not being aware of the skills and behaviors that are needed. Golfers (especially beginners) often do not realize what movements are required to perform an effective golf swing. They do not know about posture, grip, arm positions, or body rotation. Hence they have no mental representation of a movement, and so they are also incompetent at performing the movement.

Conscious incompetence
An awakening has to occur to move to conscious awareness of the behavior or skill. The player has to realize, from some source, what the swing entails. Coaches, peers, and even watching the pros on TV can create this mental shift, bringing the skill into focus for the player, who then moves from unconscious to conscious by recognizing what movements are needed. When they try to execute the skill, they soon realize they are not competent. Once this realization is internalized the player is ready to practice toward personal change and development.

Conscious competence
To get to this step, deliberate practice must take place. In this period the learner can now execute the swing with fluidity and consistency. In essence, they know how to do it and can do it. The behavior or skill does not yet happen naturally. The player must think of it, and be conscious of doing it, for the behavior to occur. But they can now do it competently.

Unconscious competence
To get to this phase, where the behavior comes naturally without thought, requires the golfer to keep on working, deliberately practicing using the skill. Eventually, the skill is used without a thought. That's the start of unconscious competence. At the end the player is proficient without thinking.

sensorimotor processing (knowing how a correct movement feels and what it looks like) to a great extent than adults who rely more on declarative (or explicit) memory (understanding the logic behind the movement). This may have the effect that juniors build stronger, longer-lasting motor patterns.

As deliberate practice continues, the variability of performance from swing to swing decreases, with the golfer being able to predict outcomes for different tactics, until eventually they master the swing movement.[3] The player better considers the

lie of the ball, target, swing, and club selection, and continues to refine their game in the second phase. The motor energy cost is lower—that is, they can hit the ball progressively farther with less effort. The golf swing becomes more or less automatic in the last phase.[1] The player does not have to attend to the details and is able to perform the swing without thinking about it. Players in this third phase are able to detect their own errors and make proper adjustments to correct them. The variability of performance becomes much smaller, and the performance improvements become gradually slower.

 **SCIENCE IN ACTION**

expect the unexpected

However much you practice, there will always be shots that do not turn out as anticipated, even if you hit the ball well. Hit exactly the same shot more than once and the outcome will differ each time. Therefore, the golfer must learn to expect the unexpected, and maintain a positive approach to their game.

Research has shown that the most successful Tour golfers are generally more mentally astute, using more consistent pre-shot routines, planning more effectively when on course, and are able to maintain high levels of confidence throughout their round.[1] Golfers constantly measure their own performance, so it can become easy for them to lose confidence in their ability in unexpected situations.

Previously, we have seen how maintaining a good mental attitude throughout a whole round is important to staying focused. Concentration, anxiety, confidence, and motivation are all key factors in effective golfing performance. If a golfer has developed a coping strategy to control these variables and bring about favorable outcomes, then that golfer will be better able to perform when unexpected events occur.

One such effective strategy is the use of mental imagery. Before each shot, Jack Nicklaus would imagine playing the shot—where he wanted the ball to land, and how it was going to get there. Even in the most awkward of positions, imagining how the swing will feel and how the shot will look can help. Specific areas of the brain are activated when we make a movement such as hitting a ball, and evidence suggests that when we imagine making the same movement very similar areas of our brain are also stimulated.[2]

Mentally rehearsing the swing and how it will feel, and visualizing the outcome, can help maintain focus, motivation, and confidence, and foster a balanced mental approach. Staying patient, accepting the bad with the good, and being prepared for the unexpected will help you to develop a positive attitude to your game.

▶ *Trunk call* *Everyone who plays golf knows that even a great shot can end up somewhere unexpected, such as next to a tree trunk, thanks to a gust of wind or an unwelcome bounce on the fairway. Learning to deal with the unexpected is an important part of playing golf well at all levels. Maintaining a positive attitude, staying focused on your next shot, and not dwelling on past shots, will all help you make the right shot decisions.*

What is the "power law of practice"?
What is the difference between learning and performance?

An ability to predict both the quantity and quality of golf practice necessary to continually improve is critical to success. Why? Because without this aptitude learning would cease and performance not progress. Performance is the behavioral act of executing a skill at a specific time in a specific situation, such as swinging a club during a golf competition. This is often visible and measurable directly. Learning, on the other hand, is a change in the capability to perform a skill. It occurs as a result of experience or practice. It cannot be observed directly, as the processes leading to change of performance are internal, with different elements of the technique learnt at different rates.[1] Investigations examining people learning refined motor skills,[2,3] similar to those used in golf, have shown that the rate of performance improvement is usually higher at the beginning and lower later on. This decrease in improvement is called a negatively accelerated function of practice, and is probably the most common feature in learning motor skills.[4]

Given that a performance improvement "slow-down" often occurs in golf, a number of plausible explanations can be deduced. The first relates to a self-identified performance proficiency—the level of playing ability a golfer is content with achieving. This typically relates to enjoyment and ability to complete the course in a reasonable number of shots. Another explanation, which is probably the more important to game improvers, is that the rate of improvement seems related to what is "left to improve." In simple terms, this means that early on in the golfer's development there is much room for advancement in skills. As the player reaches a certain level or standard, there is less room to improve and any remaining improvement may then be flatter—requiring more purposeful, refined, and committed practice. This perceived slow-down in the learning process is termed the "power law of practice."

Performance improvement trends

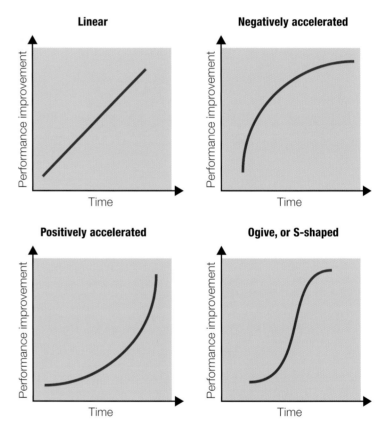

▲ **Learning curves** *In general, there are four different trends to performance improvement over a period of time: in a linear way (improving proportionally over time); negatively accelerated (improving the most in the beginning, i.e. the "law of practice");[5] positively accelerated (improving more in the end); and S-shaped or ogive (improving most in the middle). Negative acceleration in learning is affected by the time required to advance skills. A shorter time may result in a more linear progression, and a longer time in a curvilinear shape. It is important to recognize that these curves are hypothetically smoothed, with individual curves in real life likely to look more erratic. Players may experience any or all of these with different skills at any time.*

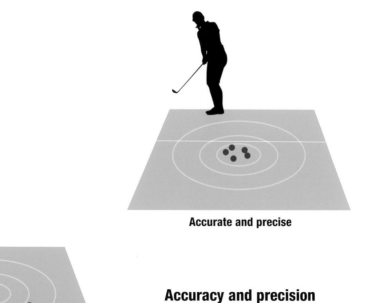

Accurate and precise

Accuracy and precision

Accurate but imprecise

Inaccurate but precise

Inaccurate and imprecise

▶ *Practice with purpose* Determining training goals helps to shape the quantity and quality of golf practice. Accuracy is measured as a central value (i.e. mean or median). For example, this may be the mean (average) distance of the group of balls to the target. On the other hand, precision (or consistency) is a measure of spread. This may be the distance between the shortest and the longest shot, or a distance measured as a standard deviation (i.e. a measure of variability around the mean). By using these measures to investigate the effects of training—for example, on impact location on the clubface, where the ball ends in relation to a target, or even golf scores—practice can become more purposeful.

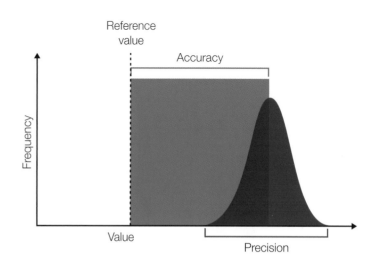

What forms of feedback improve golfer performance the most?

Should I listen to myself or to my coach?

Feedback is critical if a player wishes to improve their golf performance. Generally there are two types of feedback—task-intrinsic and augmented (or task-extrinsic). With task-intrinsic feedback, the golfer receives information from their senses. They can see, hear, and feel the movement and the outcome. This concept of feel can be divided into internal (proprioception) and tactile senses. Skilled golfers can, for example, feel the pressure from the golf club at the shaft end, and then interpret the dynamics of the shaft and adapt the swing accordingly to obtain the desired impact.[1] With augmented feedback, the player receives information from an external source such as a coach or a video camera. This provides information on which they may not have focused using their sensory system alone.

So when should a player choose augmented feedback? The decision is often based on their experience level and which aspects of their game they wish to improve. Receiving external feedback may aid the learning process and motivate the player,

allowing them to continue training toward a specific goal. It can be provided during the swing (concurrent) or after it (terminal). There are two sources of such feedback, known as knowledge of performance (KP) and knowledge of results (KR). Knowledge of performance describes how the body performs and the club is swung. Knowledge of results relates to the outcome, such as how the ball flies, how close it comes to the target, and what score is achieved. However, feedback may also hinder learning if provided incorrectly or in an untimely manner. If the coach provides too much feedback too often, the golfer may come to depend on this information and play poorly when such feedback is not available.[2, 3]

▼ **In the feedback loop** *Experienced players, through years of practice, are able to evaluate their swing themselves. They use their knowledge of performance to evaluate their movements and make corrections to improve efficiency, accuracy, and consistency. A coach must be careful how and when they provide their augmented feedback to the player, motivating them and enabling them to become an independent self-analyzer.*

> A small draw against the wind, while maintaining his balance... Nice! I'll ask him what he thinks about the swing before giving my feedback.

> Nice impact sound. That felt like a well-timed swing... and I kept my balance! Yeah... Ball is drawing nicely against the wind... I like it!

What the player hears during impact provides information on the quality of strike.

A coach should adapt feedback to suit the player.[4] A less skilled player will need feedback more frequently (every second or third swing) and in smaller units. As the player improves and the task is easier, the feedback can be less frequent (every seventh to tenth swing) and be more brief.[5]

Visual appraisal of the shot outcome by the player, such as the ball trajectory, alignment, and position on ground impact provide valuable feedback on swing effectiveness.

How the swing felt, through the body's internal proprioceptive feedback system, enables the golfer to make fine adjustments during the shot and helps with understanding the shot's outcome.

The sense of connection with the ground through the feet, and with the golf club through the hands, provides tactile feedback about the extent of pressure, strain, and balance throughout the strike.

Decision-making

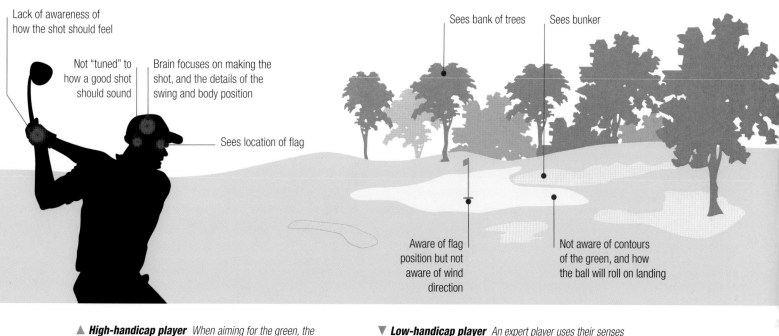

Lack of awareness of how the shot should feel

Not "tuned" to how a good shot should sound

Brain focuses on making the shot, and the details of the swing and body position

Sees location of flag

Sees bank of trees

Sees bunker

Aware of flag position but not aware of wind direction

Not aware of contours of the green, and how the ball will roll on landing

▲ *High-handicap player* *When aiming for the green, the high-handicap golfer mainly focuses on seeing the hole's position, and surrounding landmarks such as the trees at the back, and the bunker, but overall their perception of the hole is not very detailed. Their brain focuses on hitting the ball the required distance to land on the green, and because of their lack of experience and skill, this level of golfer is less attuned to how a good shot should feel.*

▼ *Low-handicap player* *An expert player uses their senses efficiently. They know almost instinctively how the shot should look, sound, and feel, freeing their brain to focus less on playing the shot and more on the details of the most relevant environmental cues—such as subtle changes in wind speed, green contours, and sandtrap positions—all leading to more accurate decision-making about the length and direction of the shot.[6]*

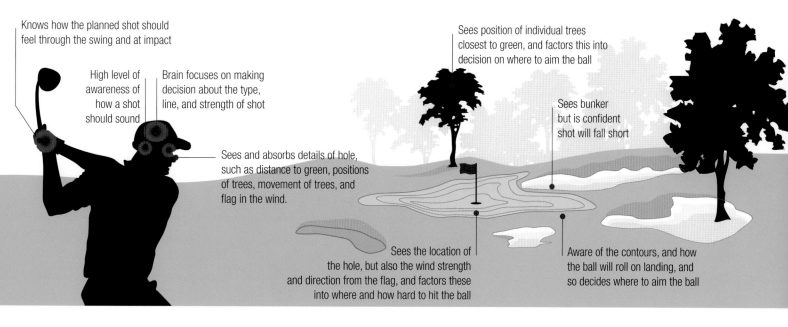

Knows how the planned shot should feel through the swing and at impact

High level of awareness of how a shot should sound

Brain focuses on making decision about the type, line, and strength of shot

Sees and absorbs details of hole, such as distance to green, positions of trees, movement of trees, and flag in the wind.

Sees position of individual trees closest to green, and factors this into decision on where to aim the ball

Sees bunker but is confident shot will fall short

Sees the location of the hole, but also the wind strength and direction from the flag, and factors these into where and how hard to hit the ball

Aware of the contours, and how the ball will roll on landing, and so decides where to aim the ball

equipment: the driver

One of the most remarkable achievements in club design is the evolution of the driver. In golf, progress usually takes decades, but in the case of the driver, this can be measured in mere years. Club designers have continually pushed the boundaries of innovation, developing the size, shape, and construction of the clubhead to maximize driver performance. The most influential changes in driver design appeared in the mid-1990s with the use of titanium, well known for its higher strength and lighter weight than steel. Designers began creating much larger clubheads, while still meeting the weight specifications of a normal driver.

This in turn had the effect of creating a higher moment of inertia and subsequently making it easier to hit the ball straight even for off-center strikes. In a short space of time, clubhead volumes developed from 190 cc to 300 cc. The year 2000 saw the first 350 cc driver, followed in 2001 with a 400 cc driver and a 500 cc driver in 2002. At this point, the rulemakers began to propose limits on drivers as they were potentially seeing technology threaten to diminish skill level. So in October 2003, the ruling authorities imposed a 460 cc limit on clubhead size from start of the 2004 season.

With restrictions of head volume introduced, club designers shifted their focus to explore shape, weighting, and material properties. Recently, attention has turned to predicting clubhead drag forces and identifying specific geometric features contributing to the total drag (pressure and friction drag) on drivers. With clubhead speeds exceeding 100 mph (161 km/h) prior to contact with the golf ball, the driver is a bulky, intrusive object which can generate significant drag force during the swing motion. The shape of the club has a definite influence on this force, and golf club designers must account for this while trying to optimize the club shape. Manufacturers' research has shown that the reduction in clubhead speed measured during player tests correlates strongly with the resulting increase in aerodynamic drag for extreme dimension clubheads (i.e. 460 cc). The use of computational fluid dynamics (CFD) has enabled engineers to make small modifications to both the face area and the transition area from the face to the body of the club to help keep airflow attached to the head surface and reduce aerodynamic drag.

▶ **Modeling airflow** *Engineers are able to model clubhead designs by applying computational fluid dynamic (CFD) technology. These techniques, often used in the aviation and automotive industries, provide cutting-edge modeling to replicate airflow dynamics around the head. The aim is to ensure the airflow does not separate during the early stages of the swing, resulting in lower drag in all orientations, and thereby improving the overall design and clubhead speed. Drivers now have increased dimensions and high inertia with low aerodynamic drag forces, enabling increased clubhead speeds and greater drive distance for golfers.*

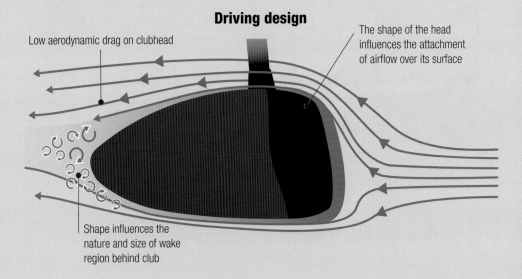

Driving design

Low aerodynamic drag on clubhead

The shape of the head influences the attachment of airflow over its surface

Shape influences the nature and size of wake region behind club

By the rule

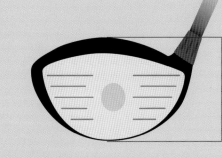

The length of the clubhead shall not be greater than 5.0 in (12.70 cm) when measured from the heel to the toe.

The volume of the clubhead shall not be greater than 460 cc. A maximum test tolerance of 10 cc is permitted.

◀ **Clubhead dimensions** *Rules governing clubhead dimensions were introduced in 2004 because of the trend toward increased driver head sizes. The Equipment Standards Committee of the Royal and Ancient determined that driver heads larger than those already permitted were not traditional and customary. Many clubheads have markings on the head to indicate approximate volume.*

The height of the clubhead shall not be greater than 2.8 in (7.11 cm) when measured from the sole of the clubhead to the crown.

Driving farther

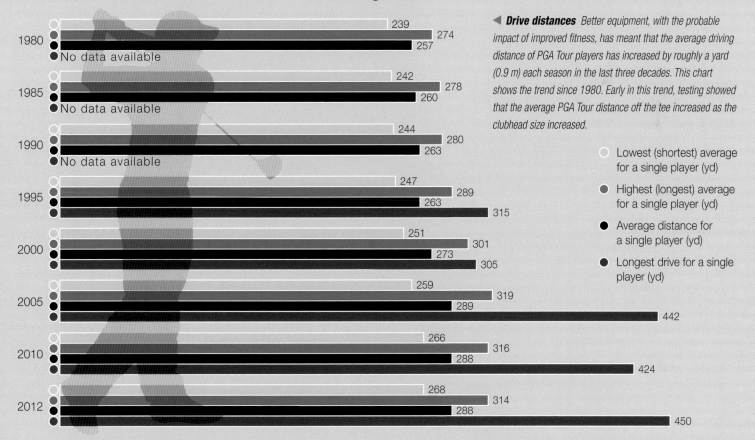

◀ **Drive distances** *Better equipment, with the probable impact of improved fitness, has meant that the average driving distance of PGA Tour players has increased by roughly a yard (0.9 m) each season in the last three decades. This chart shows the trend since 1980. Early in this trend, testing showed that the average PGA Tour distance off the tee increased as the clubhead size increased.*

1980
239
274
257
No data available

1985
242
278
260
No data available

1990
244
280
263
No data available

1995
247
289
263
315

2000
251
301
273
305

2005
259
319
289
442

2010
266
316
288
424

2012
268
314
288
450

○ Lowest (shortest) average for a single player (yd)

⬤ Highest (longest) average for a single player (yd)

● Average distance for a single player (yd)

⬤ Longest drive for a single player (yd)

How important is effective green reading to putting performance?

How should I spend my time on the practice green?

Accomplished golf performance depends on abilities in the long game, short game, and putting. Over 40 percent of strokes are taken with the putter, so time spent practicing with it should pay dividends when on course. Instructional literature on effective putting practice is often perceived as anecdotal and based on observations by top coaches and players, rather than on published scientific research. Recently however, scientists have begun to unravel the major factors that contribute to putting success, which, when deconstructed, may offer a solution as to how best to spend the time on the putting green.

Experimental research has revealed that skilled golfers have in general a five-times larger deviation in distance (6.5 percent) than in direction (1.3 percent) when making those important long putts to the cup.[1] Quite simply put, golfers tend to make more errors in how hard they hit the putt than in the line they start it at. However, mathematical calculations[2] reveal that, as long as the ball is headed for the center of the cup, it will go in even when a remarkably large distance, or range, error has been made. It will either just trickle in for a short (slow) ball or hit the back of the cup and bounce in for a long (fast) ball. The initial angular deviation from the perfect track must not be so large that the ball misses to the left or right of the hole (see captions right).

Evidence has shown that, when skilled golfers were asked to make a series of 20 ft (6 m) putts across an undulating green, the ability to effectively read the green played a more substantial role in the overall success (that is, reducing the final distance of the ball from the hole) than technique or uncontrollable green inconsistencies. With further evidence highlighting the importance of green-reading ability in the reduction of direction variability in putting, it seems logical to suggest that practice scheduling should include sufficient time to develop the skill of reading a green.[3–5] Green reading is a cognitive process, by which the player has to decide how hard

to hit the ball and in which direction to hit it. Environmental forces, such as gravity, wind, and friction (grass length and direction), affect the ball distance and direction. The ability to predict these forces, and the required clubface alignment and clubhead speed at impact, is probably the most critical factor in putting success among skilled golfers.[1] Many missed putts are wrongly attributed to poor technical proficiency, when it is more likely to be poor green reading that is the main cause of error. Thus, it is important to train with drills that give the player accurate feedback about the green conditions when hitting a putt, focusing on the most important part of putting.

Reading the green

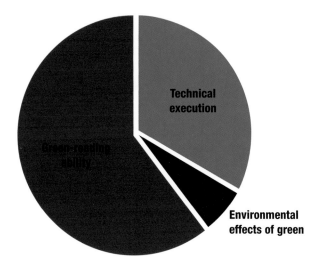

Technical execution

Green-reading ability

Environmental effects of green

▲ *Reasons for shot accuracy* *Where the ball ends up after the stroke may have more to do with a player's ability to accurately read a green than with the putting technique itself. Findings from a research study highlight that where the ball stops in relation to the target is more the consequence of green-reading ability (60 percent), than technical execution (34 percent) or environmental effects of the green (6 percent).[1,4]*

On target

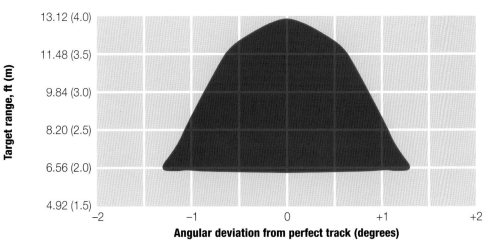

Target range, ft (m) (vertical axis)

13.12 (4.0)
11.48 (3.5)
9.84 (3.0)
8.20 (2.5)
6.56 (2.0)
4.92 (1.5)

−2   −1   0   +1   +2

Angular deviation from perfect track (degrees)

◀ **Target range for a 6.5 ft (2 m) putt** This figure shows the target range (the distance the ball would reach if uninterrupted) and "initial angle" (angular deviation) when hit that will result in the ball dropping in the cup on a sightly uphill green (of 2-degree incline).[5] This shows that for a ball approaching the hole on line (i.e. initial angle = 0º), the target range can be anywhere between 6.5 and 13 ft (2–4 m) for it to drop in. If the target range is above 13 ft (4 m) the ball will arrive at the hole too fast to go in and will skip over the hole and carry on rolling. Hit to a fraction under 6.5 ft (2 m), a tolerable angular error of ±1.3 degrees will still see the ball drop.

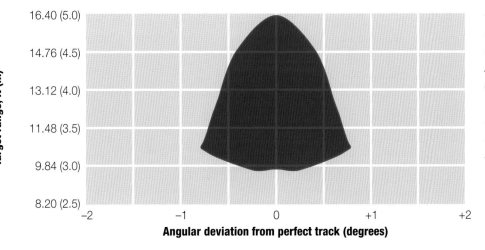

Target range, ft (m) (vertical axis)

16.40 (5.0)
14.76 (4.5)
13.12 (4.0)
11.48 (3.5)
9.84 (3.0)
8.20 (2.5)

−2   −1   0   +1   +2

Angular deviation from perfect track (degrees)

◀ **Target range for a 10 ft (3 m) putt** As the distance between the ball and hole increases, the angular deviation from the perfect track becomes more important. Again, for a slightly uphill putt, for the ball to drop into the cup the target range can be anywhere between 10 and 16.4 ft (3–5 m). However, with increased hole distance the initial angular error that still allows success is reduced to 0.75 degrees, thereby making the direction a more important factor as the length of putt increases.

Making tracks

▶ **Ball path** The track of a ball can be computed for an ideal undulating green.[5] In simple terms, the curvature of the ball's path is related simply to the local slope and to the local ball speed. On a level green the distance traveled is proportional to speed—hitting the ball twice as hard sends it four times as far.[2,5]

For a level putt, the target range is proportional to initial ball speed squared.

For a putt across an undulating green, the path of the track taken by the ball can be predicted by taking into account the slope angle and ball speed at each point.

What is the best overall pace for putts?

Should you lag your putt to die at the hole?

Given that the putt is such an important aspect of golf, the goal of attaining the smallest number of strokes possible is sometimes compromised when the ball either pulls up a fraction short of the hole, or bounds straight over it as if the hole wasn't there in the first place. Getting the speed at which the ball is captured by the hole just right can mean the difference between making par, winning a hole, a match, or, more significantly, a professional tournament. Capture speed dynamics are simple: the faster the ball is rolling at the hole, the narrower the hole effectively becomes. This is because it takes a certain amount of time for the ball to fall into the hole (i.e. 0.07 s), and if the ball is rolling too fast and/or hitting the hole off-center, there is not enough time for the middle of the ball to drop below the top of the cup. For that reason a delivery speed between one and four

revolutions per second has been recommended so that the hole width is effectively maximized and the wobble caused by the slowing ball is moved behind the hole, rather than in front of it.[1]

In examining the factors that may contribute, given a set of conditions, to the optimal delivery speed of a putt, several assumptions are made. One is the behavior of the grass at the edge of the hole, which has two effects: under the weight of the ball, the grass at the front edge of the hole will compress, allowing the ball to begin falling sooner—this means that the ball can move a bit faster when it reaches the hole and still fall in; and the grass at the back edge of the hole will undergo deformation the moment the ball strikes it, meaning that energy will be absorbed by the grass, which slows the ball, and implies

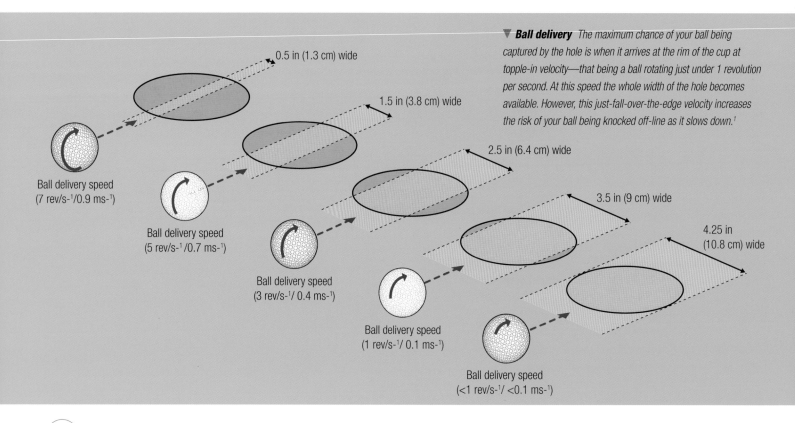

0.5 in (1.3 cm) wide

1.5 in (3.8 cm) wide

2.5 in (6.4 cm) wide

3.5 in (9 cm) wide

4.25 in (10.8 cm) wide

▼ **Ball delivery** *The maximum chance of your ball being captured by the hole is when it arrives at the rim of the cup at topple-in velocity—that being a ball rotating just under 1 revolution per second. At this speed the whole width of the hole becomes available. However, this just-fall-over-the-edge velocity increases the risk of your ball being knocked off-line as it slows down.[1]*

Ball delivery speed
(7 rev/s^{-1}/0.9 ms^{-1})

Ball delivery speed
(5 rev/s^{-1}/0.7 ms^{-1})

Ball delivery speed
(3 rev/s^{-1}/ 0.4 ms^{-1})

Ball delivery speed
(1 rev/s^{-1}/ 0.1 ms^{-1})

Ball delivery speed
(<1 rev/s^{-1}/ <0.1 ms^{-1})

that the ball can strike firmer and still fall into the cup. (It also means that the ball can hit the back edge below its equator and still drop into the hole.) Therefore the ball can arrive even faster and still fall. There is also the not-so-rare occurrence of the ball striking the back edge of the hole, popping up and falling into it. The traction of the rolling ball will play its part, too. When the ball strikes the back edge of the hole, it will still be rolling as it falls due to the conservation of angular momentum.

Appreciating the importance of delivery speed may be helpful in designing putting training programs. Players could focus on understanding the trajectory of a ball across a green, optimal ball speed and green speed, and the effects of grass friction and gravity on the ball.

Get in the hole

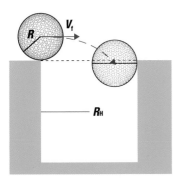

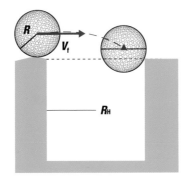

▲ *Falling in* *The mathematical problem of the capture of a golf ball by a hole of radius R_H, on a flat green has been considered in a research paper.[2] The simplest condition for the capture of a golf ball that is traveling directly toward the center of the hole is for the golf ball, after it leaves the front rim, to free-fall a distance greater than its radius, R, before it strikes the far rim, as shown above.*

▲ *Too fast* *In the simplest of terms, to sink a putt, the ball's equator must get below the lip of the cup. If the ball's final velocity (V_f) is too fast, comes in contact with an uneven surface on its way to the hole, or strikes irregularities in the turf surrounding the hole, the ball may not fall and may appear to jump out of the hole.*

Dropping the ball

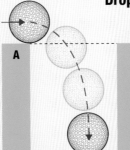

A

At the point that the ball is no longer supported by the surface of the putting green, and given the laws of gravity, it begins to fall into the hole. The moment the ball begins to fall freely, it assumes a parabolic trajectory (simplified by ignoring air resistance).

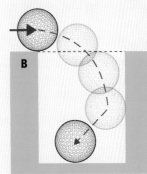

B

For a slightly firmer putt from the same distance the ball will arrive at the hole with more speed. It will follow a path illustrated here. In this case, the ball hits against the back wall of the hole before sinking to the bottom. The ball still follows a parabolic path but the parabola is broader.

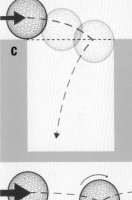

C

Presume the ball arrives at the hole even faster. It will strike the back wall of the hole higher up. Eventually, there is a speed that causes the ball to strike the back edge just at the ball's equator. Beyond this speed, the chances of the ball popping out of the hole or lipping out increase dramatically.

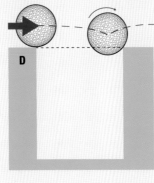

D

When the ball carries too much speed toward the hole the chances of the ball striking the back edge of the hole, popping up and falling into it are minimal—the ball is much more likely to lip out or roll right over the hole. Based on calculations, this can occur when the final velocity is greater than 10 rev/s⁻¹ (1.32 ms⁻¹). This assumes a level green and a hit at the center of the hole. Speeds will vary depending on the slope and off-center hits.[1,2]

▲ *Optimum speed* *Based on mathematical calculations there certainly appears to be an optimum speed of the ball as it arrives at the hole in order for it to drop.[2] The ball needs to have enough speed to carry all the turf irregularities, but if the speed increases over a critical value, the ball has a higher likelihood of lipping out.*

Can practice with mechanical devices enhance performance?

Will current golf training aids improve my golf?

In theory, mechanical devices that aim to control or restrict the golfer to certain postures and positions are built on sound theoretical principles and instructors, technical preferences. However, the leap of faith from theory to practice is often missing one vital link—robust empirical evidence. In practice, developing a new motor skill such as the golf swing presents a challenge to the coach and player. The intricate cognitive–neural–muscular connections require the boundaries of movement to be determined and then refined relative to the given task. The idea behind many mechanical external devices is that they provide support when establishing movement patterns. However, has experimental evidence actually proven their effectiveness in improving golf performance?

In short, the answer is currently "No." Despite claims that these devices reduce handicap by increasing automaticity, enhancing stability, developing a better swing tempo, or helping to achieve "on-plane" movement more consistently, the evidence to support these assertions is thin. Published research has shown that the use of mechanical devices in sport may have marginal effects on movement development, but not enough to actually impact on performance (e.g. lower golf score). Furthermore, it remains unclear whether the removal of the device after practice actually affects later performance—this has received very little scientific attention.[1–5] Such devices aim to act as constraints on action, either affecting the actual movement, or shaping and teaching movements.

Acting upon this type of external feedback, however, may not always facilitate learning. For example, by practicing with external feedback, such as that provided by a device, the golfer's ability to improve based on their own intrinsic feedback (for example, movement proprioception or the sound of contact) may be diminished.[6] The player may become dependent on practicing with external feedback. A device which constrains or limits movement can also alter the "feel" of the shot. Such devices also risk reducing the cognitive effort put into practicing[7] and inhibit the transfer of skill learning to new situations.[8]

▶ *Mechanical learning Mechanical devices or training aids might be useful during the initial technique development phase of learning, when the novice golfer is attempting to assemble a "ball park" movement solution.[5] A physical constraint, restricting superfluous movement of the golf club during the swing, may encourage the development of stable regions. It may also reduce attentional demands in these early stages. It is important for skilled golfers to "search and discover" their own unique movement solutions that complement their intrinsic dynamics. Training devices should then be used less frequently or not at all.*

Learning phases

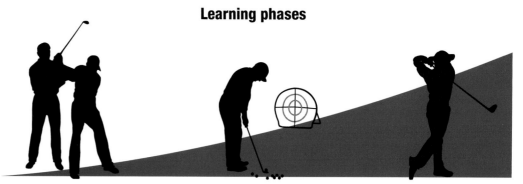

Technique development phase
A high level of focus on performing the technique correctly. High repetition phase with lower level of intensity. In this phase you practice the technique without a focus on where the golf shot finishes.

Targeting development phase
In this phase the emphasis is on hitting shots to targets and refining the shot-making process. In this phase there is almost no focus on performing the technique correctly.

Tournament transfer phase
In this phase the main focus is on the total score.

Help or hindrance?

▶ *How useful are mechanical aids?* *Based on the theoretical principles that govern the swing (i.e. the kinematic sequencing of movement), "manufacturing" or "cloning" golf swings by using artificial mechanical devices or attempting to mimic the technique of a "perfect model," respectively, may not be the most efficient or effective ways to learn to play golf. Indeed, the development of a golf swing that does not work with the unique intrinsic dynamics of the individual golfer could be prone to breaking down during play out on the course.*

Natural learning
Some devices may counteract a golfer's ability to learn

Release
Elasticated band to teach swing plane and release movements

Hinge
Aims to teach the correct wrist position at the top of the swing

Blades
Designed to guide alignment of the club during the swing

For	Against
Advocates of training aids often make claims about their effectiveness, citing that they provide feedback, can correct swing faults, focus attention, increase confidence, and improve performance.	Opposing camps claim that such devices reduce internal feedback, encourage mindless practice, are not individual-specific, encourage too much repetition of the same movement, and are often expensive.

Fins
Adds air resistance to the shaft to help build swing strength

Leg brace
Holds knee in position to restrict leg movement and build strength

Head brace
Aims to train fixed head position and arm swing movement

Arm brace
Clicking sounds indicate correct arm positions during swing

Can contextual interference aid golf practice?

How should practice be structured?

Many top-level golfers, coaches, and performance psychologists know that structured practice can have a dramatic effect on a player's rate of improvement over time. As early as the 1950s, research into the effect of altered practice structure led scientists to derive the notion of "contextual interference," which was popularized by cognitive psychologists.[1] Early findings revealed that altering the conditions in which a verbal task was practiced increased the likelihood of not only improving that task, but also remembering and transferring the skills to a novel, unfamiliar task and setting.

Contextual interference is the process of combining several tasks and practicing them together, which can be beneficial for learning.[2] When first conceived, this was quite a radical idea within the field of psychology. It gained momentum decades later when Shea and Morgan first demonstrated that the theory was valid for motor skills as well as for verbal tasks.[3] They investigated the difference in movement time when learning a skill through two forms of practice, known as "blocked" and "randomized." Practicing in a blocked manner refers to practicing the same task over and over again before starting on a new task. On the other hand, randomized practice means that task conditions are varied during the practice. After practicing a simple motor task for just 10 minutes at a time over 10 consecutive days, findings indeed revealed that randomized practice was considerably more effective in advancing the skill level of the performer. Other laboratory studies examining the performance of both beginners and skilled athletes in different sports have yielded similar findings.[4,5]

Specifically within the context of golf, research has showed mixed results. However, it should be noted that most studies have been conducted on beginners and their ability to learn how to putt. Some studies indicate that beginners when practicing in a randomized manner show greater improvement

Best practice

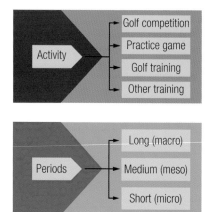

◄ *Finding the right structure*
As in any well-thought-through practice regime, consideration needs to be given to the structure, type of activity, and practice period. As optimal practice can differ depending on factors such as skill level, age, training history, and time of the year, viewing practice as short (micro) periods—such as training sessions during a week—can be a better way to manage practice over the long term.[13] Smaller chunks of practice spread out across a short timeframe is therefore recommended within the scientific literature. If a player only has time to practice once a week, then the training should probably have regular breaks and be more varied.

over time than those undergoing block practice[6–8] but other studies show no differences.[9,10] One study even found a reverse effect, with blocked practice enhancing performance over time compared with randomized practice.[11] In light of these observations one solution may be to increase the degree of contextual difficulty systematically. Thus, a beginner or a player relearning a technique may start with a block-structured practice regime. As they attain a degree of success and become more skilled, they can gradually move to a more randomized practice structure. Hitting shots from a variety of lies at different targets may not necessarily help the player improve during practice, but has the high likelihood of transferring successfully onto the course where it matters the most.[12]

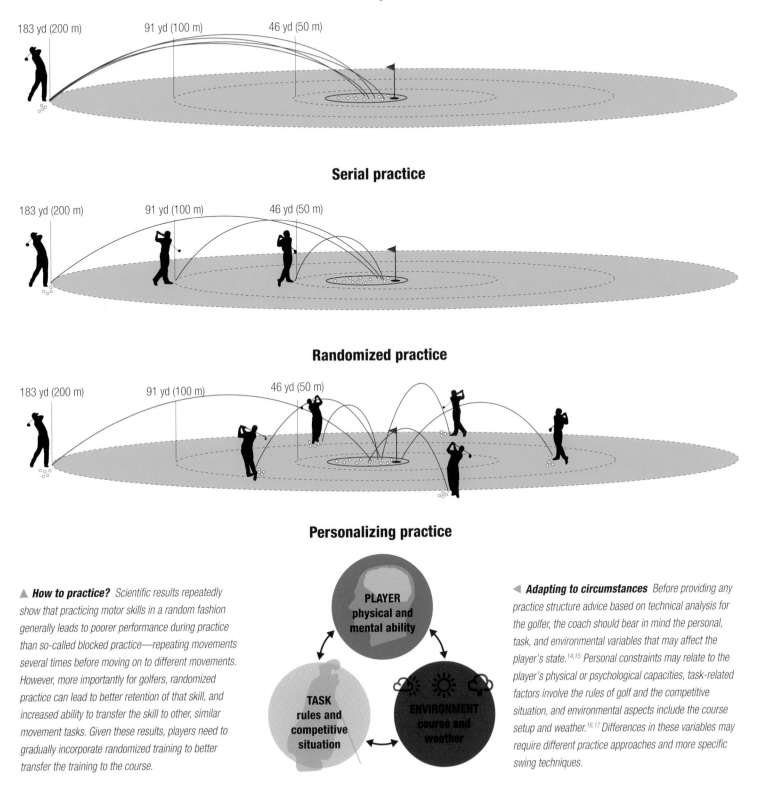

Blocked practice

183 yd (200 m)   91 yd (100 m)   46 yd (50 m)

Serial practice

183 yd (200 m)   91 yd (100 m)   46 yd (50 m)

Randomized practice

183 yd (200 m)   91 yd (100 m)   46 yd (50 m)

Personalizing practice

PLAYER
physical and
mental ability

TASK
rules and
competitive
situation

ENVIRONMENT
course and
weather

▲ **How to practice?** *Scientific results repeatedly show that practicing motor skills in a random fashion generally leads to poorer performance during practice than so-called blocked practice—repeating movements several times before moving on to different movements. However, more importantly for golfers, randomized practice can lead to better retention of that skill, and increased ability to transfer the skill to other, similar movement tasks. Given these results, players need to gradually incorporate randomized training to better transfer the training to the course.*

◀ **Adapting to circumstances** *Before providing any practice structure advice based on technical analysis for the golfer, the coach should bear in mind the personal, task, and environmental variables that may affect the player's state.[14,15] Personal constraints may relate to the player's physical or psychological capacities, task-related factors involve the rules of golf and the competitive situation, and environmental aspects include the course setup and weather.[16,17] Differences in these variables may require different practice approaches and more specific swing techniques.*

Can interactive metronome training improve golf performance?

Can listening to a rhythmic beat improve shot accuracy?

All top golfers have great rhythm throughout their entire swing. From the first movement into the backswing right through to the finishing poise, the smooth carousel of postures and coordinated body movements are all timed to perfection. Rhythm, which represents a timed movement through space, is a critical element in achieving a well-timed, coordinated movement. For over three decades, neuroscientists have explored the role of timing in the brain for such things as reading, attention, memory, cognitive processing speed, decision-making, and motor coordination. How the timing centers in the brain—which observe, control, and differentiate the rhythms of specific motor actions—work in unison has revealed insights into the complex neural synchronicity between physical and cognitive activation and functioning. Only recently has this branch of research started to explore the role internal timing may play in developing effective rhythm within the golf swing. Emerging evidence does suggest that improvements in timing and rhythmicity, augmented through listening to external auditory rhythms, can have positive effects on the outcome of golf performance.[1,2]

But what does this research tell us about how this may work? In a fascinating study first published in the journal *Cortex* in 2009,[3] researchers explored the brain under functional magnetic resonance imaging (fMRI) while subjects listened to rhythmic auditory sounds. Findings revealed that areas of the brain involved in motor planning and sequencing (or preparing motor sequences) were specifically activated while listening to rhythmic beats. Known as interactive metronome (IM) training, listening to a steady auditory rhythm seems to improve motor planning and sequencing (and thus motor coordination). Given that the golf swing requires precise timing and that a number of timing mechanisms within the brain govern such motor action, it seems plausible to suggest that the inclusion of auditory-based training can have a positive effect on improved rhythm throughout the golf swing.

▼ *Good timing* *There are a number of metronome devices on the market which help golfers to time their full swing or putting movement. Better rhythm helps the golfer to focus on the stroke and control muscle movement. Devices are available which allow golfers to adjust the number of beats per minute (typically between 60 and 70, depending on what works best for each individual), and even plug in headphones to avoid disturbing others on the driving range or practice green. The devices can be clipped to a collar or a cap.*

Swing rhythm

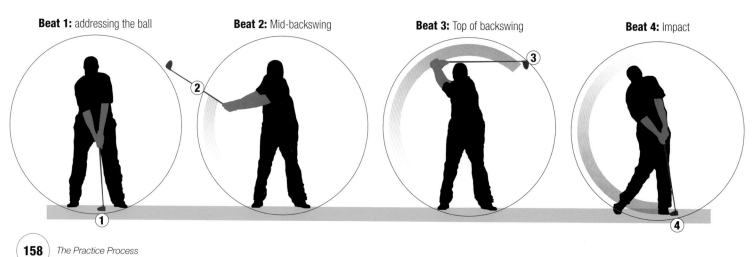

Beat 1: addressing the ball
Beat 2: Mid-backswing
Beat 3: Top of backswing
Beat 4: Impact

Sound advice

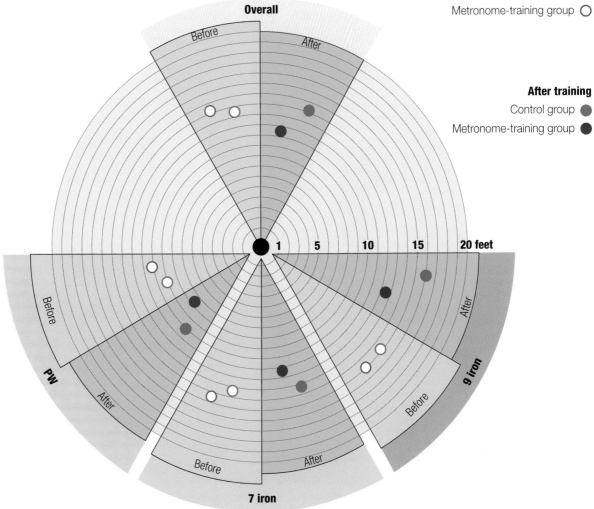

Overall

Before | After

1 5 10 15 20 feet

PW

9 iron

Before | After

7 iron

Before | After

Putt beats

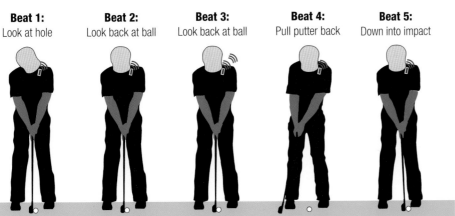

Beat 1:
Look at hole

Beat 2:
Look back at ball

Beat 3:
Look back at ball

Beat 4:
Pull putter back

Beat 5:
Down into impact

▲ ***Metronome training*** *Listening to rhythmic sounds which require the synchronization of hand and foot movements, has been shown to have positive effects on golf shot accuracy. Researchers from Sweden discovered that after 12 45-minute sessions across 4 weeks, golfers who were exposed to interactive metronome training significantly improved their golf performance.[2] Not only did they show evidence of improved motor timing throughout the swing, but there was also less variability in their ability to get close to the pin.*

159

For most players, a round of golf, whether played socially or in a competition, is over as soon as the clubs are in the back of the car, but nearly every professional and certainly 90 percent of improving golfers don't just leave it there. One thing for certain is their interest in analyzing their score to find out their strengths and weaknesses on the day and attempting to do something about it. With a growing number of players becoming interested in golf statistics as a way of highlighting game improvement areas, interest in the science behind such an approach has also gained momentum. The process of enhancing golf performance by analyzing scoring is supported by research that shows human observation and memory, wonderful though they are, are not reliable enough to provide accurate and objective information about our golf performance. Throughout this chapter, Graeme Leslie, an expert in Golf Performance Analysis, will reveal the science behind the numbers through a series of questions you've always wanted ask.

chapter seven

the score

Graeme Leslie

What are the key statistics in golf?

Which statistics will help with my game?

Golf is all about the lowest number. The key objective is to return the lowest score. Looking behind the scorecard at how the number is composed, the performance objectives of golf can be reduced to two things—hitting greens and holing putts. The golfers who are better than average at both finding GIRs (greens in regulation) and at PpR (putts per round) will score better than most of their peers. Often, a golfer who is below average in either one of these two performance stats will score better than average. This will be due to exceptional ability in either the long game or putting skills.

The GIR and PpR statistics are linked, however. When a golfer hits considerably more greens than their playing partner, the average length of the first putt is usually longer. When a golfer misses a green in regulation, the shot on to the green will inevitably be from much closer than the distance from which an approach shot was made. The pitch, chip, lob, flop, or bunker shot, or even the putt from off the green, will result in a closer proximity to the hole than the average approach shot, thus providing better opportunities to single putt. The number of putts per round is therefore heavily dependent statistically on

Reaching the green

▶ *Greens in regulation*

Finding a green in regulation consists of one (tee) shot at a par 3, a drive and an approach at a par 4 (and often the same at par 5s when the green can be hit in two), and at longer par 5s, a drive followed by a positional shot followed by an approach (or a short game shot) for the third shot. Strong drivers and weaker iron players can hit the same number of GIRs as weak drivers and strong iron players, only they find different parts of the course in getting there.

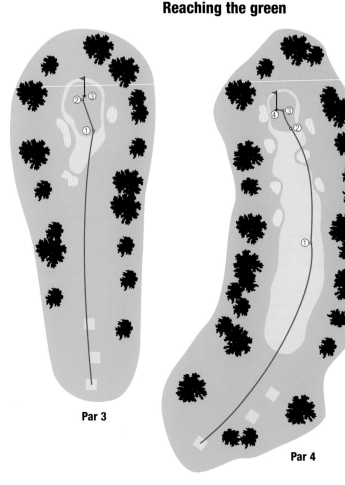

Par 3

Par 4

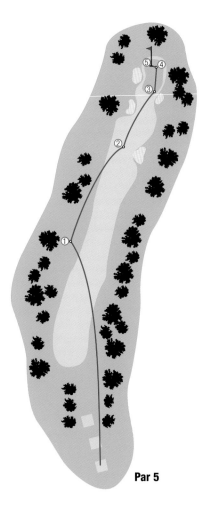

Par 5

Top of the Tour pros

	Greens in regulation			Putts per round		Average score for year
	Per round	%	Position on tour	Per round	Position on tour	
Lorenzo GAGLI	13.9	77.1	3	31.3	198	70.54
Richie RAMSAY	13.7	75.9	7	30.9	190	70.60
Thaworn WIRATCHANT	11.5	63.9	172	28.6	4	70.68
Nicolas COLSAERTS	13.6	75.4	9	31.0	193	70.69
Stephen GALLACHER	13.2	73.5	18	30.7	180	70.85
Chapchai NIRAT	13.8	76.4	5	31.5	201	71.00
Justin ROSE	13.9	77.2	2	30.7	179	71.00
Tetsuji HIRATSUKA	11.2	62.5	183	28.4	2	71.28
Soren HANSEN	13.0	72.2	27	30.7	192	71.36
Ian POULTER	13.5	75.2	11	30.6	172	71.40
Gary BOYD	12.9	71.9	33	31.1	195	71.51
David HOWELL	11.2	62.5	184	28.7	5	71.70
Carlos DEL MORAL	11.3	62.7	180	28.9	9	71.72
Steven O'HARA	13.3	73.6	17	31.4	200	72.01
Sam HUTSBY	11.1	61.4	191	29.0	16	72.65
Jose Maria OLAZABAL	10.6	59.1	198	29.2	29	72.78
S.S.P. CHOWRASIA	10.6	58.8	199	29.0	15	72.90
Daniel VANCSIK	10.9	60.4	194	29.1	17	73.21
Mark F. HAASTRUP	10.4	57.8	202	28.8	8	73.44
Jarmo SANDELIN	10.5	58.6	200	29.0	12	73.44

the length of the first putt, and the more greens that are missed, the more opportunities there are to pitch it close and single putt, usually to save par.

GIR stats are made up of a combination of different shot types. The par 3 tee shot that finds the green is a one-shot GIR whereas at par 4s, the drive and the second shot—an approach usually but sometimes a short game shot—are required. At particularly short par 4s, or where there is a howling gale downwind or fast-running links turf (as found in the UK in summer), helping the drive to extraordinary lengths, the second shot to the green resulting in a GIR can be from

▲ **Analyzing the Tour pros** *This table shows a number of players on the European Tour 2011, their GIRs and PpRs, their positions on Tour in these two measurements, and their score averages. All these examples show a clear inversely proportional relationship between GIR and PpR. On average, the more greens found (good), the more putts are taken (bad), and the fewer greens found (bad), the fewer the number of putts (good).*

greenside or close to it rather than a full approach. The shorter par 5s may be found in only two shots but at the longer ones where the golfer cannot reach in two, the drive will be followed by a positional shot, with the third shot being required to hit the green for the GIR, whether a full approach or a short game shot.

Which areas of the game most influence the scorecard?

What should I work on to improve my scoring?

The interdependent nature of golf means that the relative difficulty or ease of any particular shot is determined by the quality of the golf shot that precedes it. The length of the second putt will depend on the proximity of the first putt. The first putt's proximity to the hole is decided by the quality of the bunker, pitch, or approach shot that put the ball on the green. At a par 4 hole, both the degree of difficulty and the length of the second shot are determined by the quality of the drive.

There has been much examination of what constitutes the "most important" part of the game. One study at Columbia University[1] concluded that "long game shots (those starting over 100 yd (91 m) from the hole) explain about two-thirds of the variability in scores among golfers on the PGA Tour." Another study[2] of all the winners (and tied firsts) on the European Tour for the decade 2000–2009 concluded that the improvement in putting performance was greater than the improvement in greens in regulation for golfers when winning. From the first million shots recorded on the Golf Data Lab database between 2009 and 2012, by professionals and elite amateurs all over the world, the "two-thirds long game" conclusion seems statistically improbable. On the contrary, the short game and putting—as well as the approaches from less than 100 yd (91 m)—appear to have the bigger effect on scoring variations. It is known that 50 percent of the game is played from 15 yd (14 m) or less from the hole, with putts on the greens alone accounting for 42 percent of the total score.

In order to deal with the question of which parts of the game are the most influential, each golfer must examine their own games and determine where their own performance areas are strong or weak. No two golfers share the same performance statistics nor the exact same strengths and weaknesses. It is only by properly measuring golf performance that one can learn where to focus in practice to improve scoring averages.

Tour pro shot composition

Shot types	Total	per round
Drive	1279	14.1
Par 3 tee shot	364	4.0
Positional / recovery	94	1.0
Approach > 250 yd	49	0.5
Approach 230–250 yd	65	0.7
Approach 200–229 yd	85	0.9
Approach 180–199 yd	98	1.1
Approach 160–179 yd	139	1.5
Approach 140–159 yd	143	1.6
Approach 120–139 yd	228	2.5
Approach 100–119 yd	141	1.5
Approach 75–99 yd	180	2.0
Approach 50–74 yd	50	0.5
Short game 30–49 yd	48	0.5
Short game 20–29 yd	42	0.5
Short game 15–19 yd	32	0.4
Short game 10–14 yd	95	1.0
Short game < 10 yd	287	3.2
Putts from off	48	0.6
Putts on	2912	32.0
Penalty shots	11	0.1
Total	**6390**	**70.22**

▲ **How to use scoring data** *The data from this particular professional for 91 rounds recorded on Golf Data Lab in 2011 show an excellent average score of 70.22 and the composition of each type of shot that makes up the score. Of his score, 45.6 percent was made up of putts on the green, an extraordinarily high proportion. Further detailed analysis of his putting conversions revealed that his 32 putts per round did represent a significant opportunity for improvement.*

Critical zone

◀ **Area of focus** At professional and elite amateur level, just over 50 percent of the score on average consists of shots played from within 15 yd (14 m) of the hole. Adding up the putts on the green and all short game shots played from 15 yd of the hole and closer, including putts from off the green, bunker shots, all pitches, chip shots, lobs, and flop shots, the total of these shots alone exceeds 50 percent of the score in most cases.

The whole game

Fairway

Putting

Driving

Chipping

Bunker

◀ **Analyzing a player's game**
By employing statistical analysis, it can be found that two golfers may have the same (or very similar) average score but one needs to improve their ball-striking while the other struggles with their putting. Individual strengths and weaknesses can only be determined using a proper measurement framework.

What are the actual objectives of each shot type?

What should I be targeting and trying to achieve on the course?

Different shot types in golf have specific objectives. The primary purpose of the drive, for example, is to avoid trouble and to place the ball in a position where an approach attempt can be made; whereas the objective of every 3–5 ft (1–1.5 m) putt is to hole it. The positive and extremely confident bunker player may be trying to hole every shot from sand, but for the majority of golfers, simply getting the ball as close to the hole as possible will be the purpose of most greenside bunker shots. The recovery shot objective from deep rough may simply be to advance the ball to a better or clearer position, or, more often, to get the ball back in play, find the fairway, and create the opportunity for finding the green with the subsequent shot.

In professional and elite amateur data from 2009 onward,[1] one of the biggest surprises is how often the green is missed with short approaches, using the 9 irons and various wedges. Granted, not every start lie is from the fairway and semi-rough, as some come from bunkers and rough, but more than 1 in 10 shots fail to find the greens on average, even from good positions. Analysis of all 9-iron and wedge approaches has shown that more than 15 percent, on average, fail to find the putting surface. Statistically, just over half of all approaches are short—played with the 9 iron and wedges—so this shot type is very important to overall success.

The primary objective of every approach, including the short approach, is to find the green. When the pin is cut close to the edge of the green—for instance, tight front left with bunkers guarding front and left—then it is even more important to leave a putt for birdie, of whatever length of first putt, rather than having to work to take two from off the green to save par by single-putting. Eliminating bogey possibilities is the first objective, and, as scoring and success are mostly about reducing the incidence of dropped shots rather than birdie blitzes, a more conservative short approach targeting

8–12 ft (2.5–3.5 m) right of the pin with a knock-down gap wedge from 70 yd (64 m), rather than straight at it, usually takes a bogey right out of the equation, while still creating a birdie opportunity if a single putt follows.

Shot types

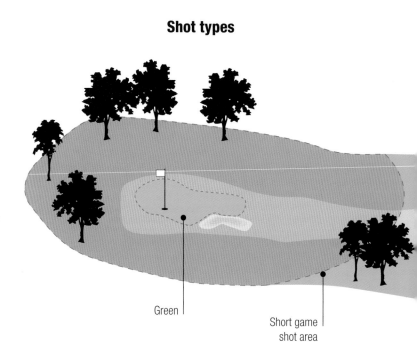

Green

Short game shot area

▲ *Course management* *There are only three types of shot that make up the "long game" that determine whether the green in regulation gets found or not (four if we include the relatively minor statistical impact of the positional second shot at the longer par 5s). These are the drive, the approach, and the par 3 tee shot. Planning each shot carefully, based on ability and environmental conditions, is critical to improving your score. One European Tour player recorded over 100 rounds with Golf Data Lab and found that his short approaches—approaches from all lies played with 9 irons and wedges—were only hitting the green two-thirds (66.78 percent) of the time. These were all genuine approach attempts rather than positional or recovery shots. On learning this, he and his coach were*

Putting target

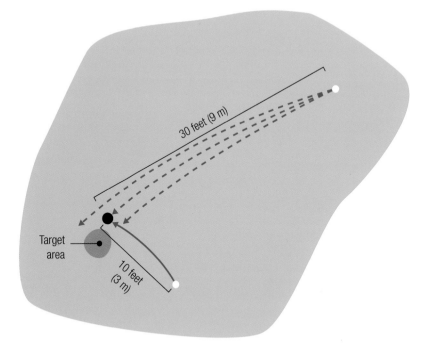

▶ *Putting objective* *From 10 ft (3 m), every golfer's only objective should be to hole the putt. From 30 ft (9 m), many of the most confident putters will also be trying to hole it but many will be trying not to three-putt as their principal objective. Instead they will be seeking to place the ball within a close enough distance to guarantee the two-putt. In this case, they might visualize the putt result to get within an area such as this, following the break of the putt but still giving the putt enough distance for the chance to drop.*

Drive

Approach
from here

able to determine that this had been a course-management issue rather than anything technical. Knowing that he had an excellent short game and putting (which was proved by his specific data), he had been too aggressive and felt that if he was to miss the green, he would get up and down to save par anyway. It was the birdie opportunities that he was missing out on. By understanding that he holed more often per round with putts from between 10 and 20 ft (3–6 m) than he holed out from off the green, he learned to take more conservative lines to tightly cut pins, to much improved effect.

SCIENCE IN ACTION planning a strategy

Even if you are a casual golfer, and not trying to win a tournament, you will enjoy your golf much more if you have "a plan." Every time you set out on a golf course, having a strategy for playing each hole and the course will help your confidence and your score.

Whether judging the wind or the grain of the grass, whether playing it safe or going for the green, the golfer is constantly assessing a whole range of factors during a round. With so many variables to consider, success in golf is heavily dependent on strategic thinking.

All the truly great players share exceptional abilities in strategic thought. As part of their planning process, before they even play, they are able to create stories or narratives set in the future which explore how their game would change if certain trends were to strengthen or diminish, or various possible events were to occur. Playing through these scenarios can be used to review or test a range of plans, the conclusion generally being that different plans are likely to work better in different scenarios. Alternatively, these imagined scenarios can be used to stimulate the development of new plan, or form the basis for a strategic vision of change.

At any given point in time, there are an infinite number of possible scenarios. Scenario planning does not attempt to predict which of these will occur, but—through a formal process—identifies a limited set of examples of possible on-course futures. These imagined outcomes provide a valuable point of reference for the golfer when evaluating current strategies or formulating new ones. Adopting a scenario-based strategy before you even step out to play—considering how you will feel and behave, the course layout, the weather, and the competition—will help you to take each shot one at a time, avoid letting the unexpected derail your plan or your confidence, and ensure you stay focused and calm.

▶ *Thinking ahead* Peter Tomasulo and his caddy assess their approach to the hole ahead from the 13th tee at McKinney, Texas. "Perfect planning prevents pathetic performance" may be a cliché but it's particularly true for playing any sport, including golf. Knowing your game—and your strengths and weaknesses—and then planning accordingly how you will tackle each hole and the course in advance of your round will pay dividends.

What do the "sand saves" and "scrambles" statistics tell us?

Is measuring getting up and down from the sand really that important?

The different shot types in golf are distinct yet interdependent in that the degree of difficulty of one shot, and the overall effect on the score, will depend on the quality of the previous one. So is there any merit in combining stats rather than simply measuring distinct shot types on their own? The mantra used by most golf psychologists is "one shot at a time." This is very good advice but easier said than done. As an aid toward achieving this, recognizing "one shot type at a time" helps to understand the distinct objectives of each individual shot in golf. For example, the greenside bunker shot objective will be to put the ball as close to the hole as possible for a single-putt opportunity.

A very positive and confident bunker player may even be trying to hole the bunker shot as the primary objective. However, a confident bunker player may not be a confident putter and vice versa. Some excellent short-game proponents are poor putters. By combining these two distinct shot types—short game and putting—and measuring "up and downs," we learn nothing of the relative strengths and weaknesses of an individual golfer. Separating bunker shots from all other short game shots is useful to determine "bunker shot-ability," but only when the

results of the individual shots are collected. The traditional "sand save" statistic cannot possibly inform us how effective the golfer is when playing from sand. As the percentage of "sand saves"—over a tournament, a month, or a whole season—combines the bunker shots and the putts, the danger is that the poor bunker player and excellent putter may have the same "up and down" percentage as the excellent bunker player and poor putter. One superior bunker player may put the ball to an average of 7 ft (2.1 m) from greenside bunkers when the ball is starting less than 30 ft (9.1 m) from the hole, whereas another may average 14 ft (4.2 m). This statistic alone informs us of "bunker shot-ability" from this starting distance. Adding whether the putt was converted to the measurement criteria or not confuses the exercise. Of course it is important to get up and down from around the greens, including from sand; this helps keep the total score down. However, in terms of understanding individual strengths and weaknesses, combining shots and shot types can never inform the golfer as effectively as measuring individual and separate shot types. Putting ability is one skill, short game another, with bunker shots a separate and distinct subset of the short game.

Sand saves

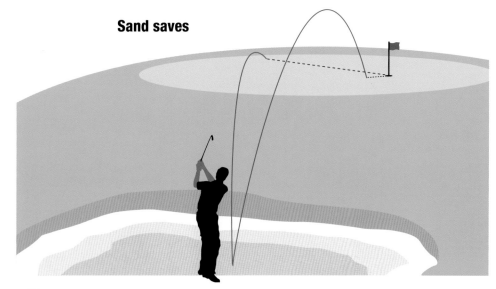

◀ ***Same result*** *The golfer may splash out to 2 ft (0.6 m) one day whereas faced with the same shot the next day from the same bunker, slightly thins it 16 ft (4.9 m) beyond the hole. When both first putts are holed, both count as successful sand saves. The fact that one bunker shot got very close and required a tap-in whereas the other was 16 ft away and needed a long range putt is not recognized in the traditional sand save stat. The combined stat provides no information on the quality of the bunker shots or the conversion ratios of specific distance putting.*

Bunker mentality

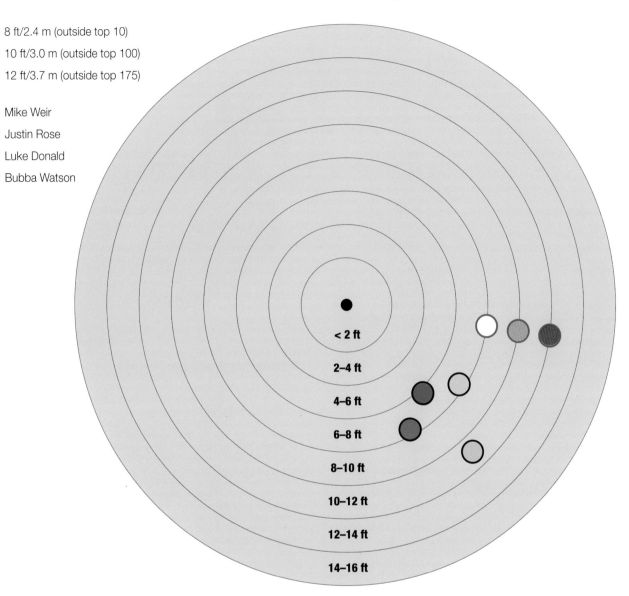

Legend:
- 8 ft/2.4 m (outside top 10)
- 10 ft/3.0 m (outside top 100)
- 12 ft/3.7 m (outside top 175)

- Mike Weir
- Justin Rose
- Luke Donald
- Bubba Watson

Radial rings (center outward): < 2 ft, 2–4 ft, 4–6 ft, 6–8 ft, 8–10 ft, 10–12 ft, 12–14 ft, 14–16 ft

▲ **Top shots** Luke Donald's recent reputation as one of the best bunker players in the world has been earned. He was tied 2nd with Justin Rose in 2011 at 7 ft 6 in (2.3 m) and tied 2nd with Padraig Harrington in 2010 at exactly 7 ft (2.1 m)—the proximity to hole from sand. From an examination of a 75,935-round data sample on the US PGA Tour for the full years 2007 to 2011, Mike Weir of Canada was clearly the best bunker player, however. In 2010, his average 5 ft 9 in (1.7 m) was the closest average of any golfer over the period. Despite Mike Weir being more than a foot closer on average from bunkers than Luke Donald in 2010, he had a "scrambling" ranking of 82nd, whereas Donald was 4th best that year. Donald was clearly much better at putting that year. This is a clear example of how the "up and down" scrambling percentage fails to describe the real picture, with Weir being clearly better from bunkers and Donald being the much better putter (1st on "strokes gained" measurement as against 107th for Weir). Bubba Watson, however, the 2011 Masters Champion, has been averaging 10 ft, 11 in for his 410 rounds during these five years, a performance from sand that has earned him an average position of 153rd best bunker player.

How effective is the Driving Accuracy statistic?

> How do I measure my driving?

The drive is one of the seven distinct shot types in golf. Per round, to the nearest whole number, there are 14 drives on average. Driving statistics for the European and US Tours cumulatively track the number of fairways hit with the drive in each round. This measure of fairways hit or missed is termed the "driving accuracy" and this is used as the principal indicator of driving effectiveness. A drive is classified as being the first shot at all par 4s and 5s irrespective of the club used.

Drive results can be one of seven possibilities: fairway, semi-rough (or first cut), bunkers, rough, position Z (e.g. the middle of trees or gorse), hazards, or out of bounds. Data collected from the European Tour to elite amateur level show that the proportion of drive results which find the semi-rough, left or right of the fairway, can be as high as 35 percent, particularly where more links courses are played, and the average percentage is around 20 percent.

As the purpose of the shot type known as the drive is to find a position on the hole from where a successful approach may be attempted and the green in regulation found, finding the fairway itself is a secondary objective. With such a significant proportion of drive results missing the fairways and yet still providing positions from which a successful approach can be made, the traditional driving accuracy statistic is not a particularly useful guide to overall driving effectiveness.

However, the incidence of drive results that find a position from which a successful approach can not be made is between 10 and 20 percent for the most effective drivers. The least accurate drivers will find the rough, bunkers, or worse with over 20 percent of their drives, sometimes more than 25 percent and occasionally more than 30 percent. This range of 10–30 percent of inaccurate drives resulting in "undesirable" positions provides a far more accurate illustration of overall driving effectiveness than the traditional measure of fairways hit or missed.

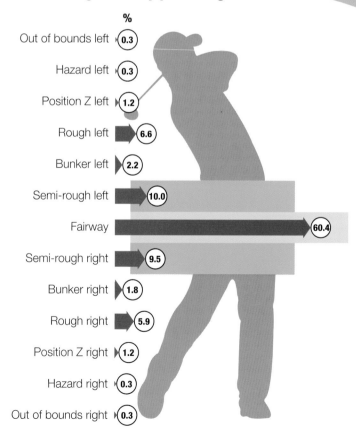

Driving accuracy percentage

%

Out of bounds left	0.3
Hazard left	0.3
Position Z left	1.2
Rough left	6.6
Bunker left	2.2
Semi-rough left	10.0
Fairway	60.4
Semi-rough right	9.5
Bunker right	1.8
Rough right	5.9
Position Z right	1.2
Hazard right	0.3
Out of bounds right	0.3

Reaching the green

◀▶ ***Margin for error*** *On this drive, the golfer with good course-management skills will be targeting the right half/right edge of the fairway to avoid the hazard on the left, knowing that if the ball goes slightly left of the intended target line, they will still be playing the approach from the fairway, and if slightly right, they will be playing from the right semi, the first cut on the right from where they know that they can still find the green with an approach shot. On particularly tight fairways with a hazard threatening on one side, the golfer may even be targeting the semi-rough on the opposite side, ignoring the fairway altogether.*

◀ ***Fair play*** *In 2011, 60.4 percent of the drives by Scotland's best amateurs (+2 handicap and better) found the fairways. Almost half of the remaining 40 percent (19.5 percent) found the left or right semi-rough. The green may still be reached from most of these non-fairway positions, including from the rough, and often from sandtraps.*

equipment: the putter

If there is one piece of equipment that has most captured the imaginations of golf club designers, golf enthusiasts, and the rulemakers, it's the putter. From the myriad of mallet and heel–toe designs to the wide spectrum of handle and grip configurations, today's putters feature the latest technology to allow golfers to improve their scores. Despite three of four consecutive majors having been won by golfers using the longer belly putters, there is no statistical or biomechanical evidence to show that the longer putters are more effective than conventional putters. On the contrary, 99 of the best 100 putters on the US PGA Tour 2011 were not using longer putters. This may become academic, however, as the R&A and USGA have now ruled that, from 2016 onward, putters which anchor to the body will be outlawed.

Keegan Bradley won the 2011 US PGA at Atlanta Athletic Club, the first winner of a major using something other than the traditional "short" putter. In 2012, Webb Simpson won the US Open and Ernie Els won The Open Championship at Royal Lytham and St Annes (beating Adam Scott, who himself uses a long putter, by one shot), both winners also using the longer "belly" putters. From no winners in the history of the game until August 2011, three winners of four consecutive majors—Bubba Watson being the exception at The Masters 2012—used the extended putter versions. This is currently a "hot topic" in the world of golf and the ban on anchored putting by the R&A and USGA has generated much discussion. The use of "fixed anchor" longer putters is at its all-time highest on tour and it is understood that some colleges in the USA are actively encouraging their golfers to try them. The game had always traditionally been about the hands on the grip of the club being

A better putter?

▶ **Unfair advantage?** Does the belly putter provide an unfair advantage? Biomechanically at least, this question has yet to be fully answered. Based on biomechanical principles of golf putting, it is reasonable to conclude that the belly putter does not provide an advantage because of its anchoring point. Since the putter is rotating about a different axis than the swinging arms, more degrees of freedom are introduced, which is not ideal for a precision task with a relatively limited range of motion. With the conventional putter, the arms, hands, and putter move nearly as one unit; however, when using a belly putter the angle between the forearms and the putter shaft must change considerably, which probably outweighs the potential advantage of anchoring the putter against either the chest or stomach.

A With a normal putter, the prominent technique taught is to reduce the degrees of freedom (DOF) so that, ideally, the putter-arm unit operates with only one degree of freedom, unlike the belly putter.

B Belly putters are about 6 in (152 mm) to 8 in (203 mm) longer than a normal putter and are designed to be "anchored" against the stomach of the player. This supposedly reduces the importance of the hands, wrists, elbows, and shoulders.

C A long putter is longer still. It is designed to be anchored from the chest or even the chin and similarly reduces the impact of the hands, wrists, elbows, and shoulders. The switch to longer putters is likely to reduce the "feel," which, for many players, is a source of error.

the only contact between golfer and equipment. The swinging motion of the club or putter head was produced with only the hands being in contact with the grip of the club and it was this motion that projected the ball toward the target. Anchoring a putter against the chin, the chest, the belly, or even the left forearm represents a fundamental difference in how the shots are played today compared with traditionally.

An analysis of the stats suggests that the increased use of belly and longer putters does not confer any distinct advantage, however. By examining the actual conversion ratios of the golfers on the US PGA Tour 2011, only Webb Simpson was inside the top 100 best putters and therefore 99 of the best 100 putters were using the conventional "short stick." The US Tour "total putting" stats showed Webb Simpson as 40th best in 2011, with Keegan Bradley being 110th and Ernie Els 180th. Even when winning the Open in 2012, Els hit 57 of 72 greens, three more than Luke Donald, who hit the second most greens that week. Els won the Open with superior ball-striking, not putting. The debate concerning long putters will inevitably rage on, but there does not appear to be any advantage from their increased use, to the end of 2011 at least. Recent Tour winners have simply brought the ongoing debate into sharp focus.

Putting factors

▲ **Putting analysis** *Putting performance can be broken down into a number of key factors, each having a determining effect on the finishing position of the ball.*

▼ **Putting accuracy** *As with any shot in golf, there are many factors to be taken into account when assessing a golfer's putting stroke. One of these is controlling the angle of the putter head and its direction on impact with the ball.*

Take aim

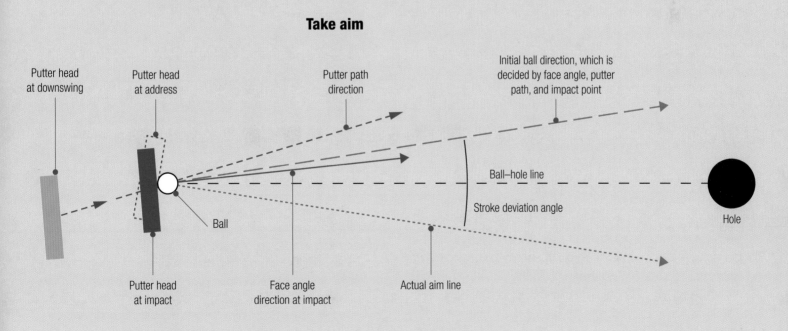

How important are putting statistics to overall scoring success?

Does better putting or better ball-striking contribute more to winning?

From making the first strike off the tee to that final roll into the cup, the golfer encounters a myriad of different shot types throughout their round—the drive, layup, approach, recovery, bunker shot, chip, pitch, and putt, to name a few. By recording these shots a picture emerges that can reveal which shot measurement, or metric, may have the most impact on overall scoring success.

Greens in regulation (GIR) provides a good indication of the effectiveness of the long game (tee-to-green) while putts per round (PpR) provides a generic picture of putting efficiency over the course of a tournament, a month, a season etc. An examination by Golf Data Lab of over 400 winners (and tied firsts) on the European Tour in the decade 2000–2009 (excluding majors and WGC events) compared the performance of Tour winners during their winning rounds against their performance during the rest of that season. Across a decade's worth of data, players were found to hit an average of 75.2 percent of GIR in events which they won, compared with an average of 69.3 percent GIR across their entire decade. This difference is the equivalent of just over one more green per round. The average number of putts they took when winning an event was calculated to be 28.34 PpR, more than one putt per round lower than they normally took: 29.48 PpR.

Each year, the European Tour maintains a table in which each golfer's actual performance is tracked against his peers. By plotting the average GIR and PpR stats of the winner against their season-long averages in the years that they won, and examining these numbers against the annual performances of the whole Tour, the "position on Tour" of the specific performance measurements can be established. For GIR, the average greens found for the winners over the whole decade improved from a position that equated to 40th best on Tour (40.2) to an average of 4th on Tour (4.4) when winning events. For PpR, the winner's average position improved from 62nd best (62.1) on tour for the

decade to 4th (3.7) during the weeks that they won. This bigger improvement—58 places for putting compared with a 36 places for GIR—is conclusive evidence that players, in order to win, need to putt better than they normally do more than they need to find more greens.

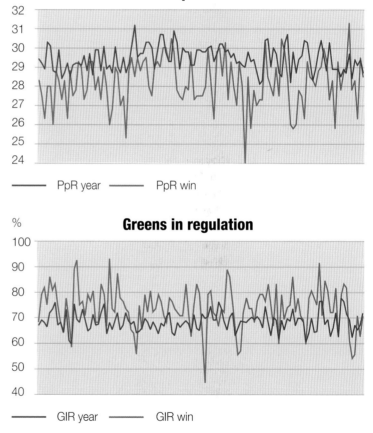

Putts per round

— PpR year — PpR win

Greens in regulation

— GIR year — GIR win

▲ *Stat attack* *By unraveling patterns in winners' performances, we can start to reveal what's important to overall success. Analysis of data between 2000 and 2009 (also examined for 2010, 2011, and 2012) tells us that when European Tour players won a tournament (i.e. when winning) both their collective GIR and PpR performances were consistently better than their overall yearly averages (i.e. having been a winner during that year). As we can expect, winners take fewer putts and hit more greens than they normally do, when winning.*

Game changers

	When winning				For whole year			
	Greens in regulation		Putts per round		Greens in regulation		Putts per round	
	Average %	Position	Average	Position	Average %	Position	Average	Position
2000	76.6	3	28.38	1	70.4	29	29.53	54
2001	77.0	4	28.49	8	71.0	30	29.54	85
2002	73.1	3	27.96	2	67.5	43	29.20	58
2003	75.5	5	28.37	2	69.4	46	29.31	53
2004	75.6	2	28.16	1	68.6	41	29.40	59
2005	74.1	4	28.56	10	68.7	28	29.34	52
2006	73.6	7	28.25	4	68.8	42	29.34	61
2007	73/7	5	28.39	3	68.4	58	29.65	62
2008	75.9	1	28.39	4	68.9	45	29.71	58
2009	75.7	10	28.42	2	71.0	40	29.75	79
Averages	**75.2**	**4.4**	**28.34**	**3.7**	**69.3**	**40.2**	**29.48**	**62.1**

▲ **Relative improvements** This table shows the relative improvements in the positions on European Tour rankings for the difference in GIR and PpR between tournament winners and their averages for the year. Comparing these difference scores (GiRdiff and PpRdiff) between 2000 and 2009 reveals that every year without exception improvements in putting difference exceeded improvements in GIR difference. For example, in 2009, an improvement of 30 places was observed in GIRdiff, while an improvement of 77 places was observed in PpRdiff. These key stats again show that, time after time, the capability to reduce the number of putts per round has a major influence on a player's ability to win a tournament.

Winning ways

Rory McIlroy	Position on Tour			
	2009	2010	2011	2012
Strokes gained	150	145	130	82
5–10 ft (1.5–3 m)	165	167	175	71
5–15 ft (1.5–4.5 m)	161	139	155	76
5–10 ft conversions	51.0%	50.6%	50.0%	57.3%

◄ **Rory on top** Rory McIlroy's superb season in 2012 cemented his place as world number one and concluded with victory in Dubai, where he putted excellently, finishing the tournament with five straight birdies, as best in putts per GIR, and with the second-fewest putts of the week, one behind Jimenez who found eight fewer greens. This table shows McIlroy's position on tour relative to the rest of the US PGA Tour pros over the last four years and his actual putting conversion ratios from the key 5–10 ft (1.5–3 m) distance. His improvement in 2012 is clear, and from not exceeding 51 percent of all putts faced between 5 and 10 ft previously, he raised this bar to 57.3 percent in 2012.

Notes

CHAPTER 1 mind and body

PAGES 14–15

1. T. C. Sell, J. P. Abt, and S. M. Lephart (2008) "Physical activity-related benefits of walking during golf," in D. Crews and R. Lutz (eds), *Science and Golf V: Proceedings of the World Scientific Congress of Golf,* Energy in Motion, Mesa, AZ, pp 128–132.

2. C. Thompson, K. M. Cobb, and J. Blackwell (2007) "Functional training improves club head speed and functional fitness in older golfers," *Journal of Strength and Conditioning Research,* 21, 131–137.

3. M. F. Smith (2010) "The role of physiology in the development of golf performance," *Sports Medicine,* 40 (8).

4. S. M. Lephart, J. M. Smoliga, J. B. Myers et al (2007) "An eight-week golf-specific exercise program improves physical characteristics, swing mechanics, and golf performance in recreational golfers," *Journal of Strength and Conditioning Research,* 21, 860–869.

5. M. F. Smith, A. J. Newell, and M. R. Baker (2012) "Effect of acute mild dehydration on cognitive-motor performance in golf," *Journal of Strength and Conditioning Research,* Nov; 26 (11): 3075–80

PAGES 16–17

1. J. N. Vickers (1992) "Gaze control in golf putting," *Perception,* 21, 117–132.

2. J. N. Vickers and D. Crews (2002) "Short-term memory characteristics of golfers: Findings from concurrent measurement of gaze and EEG research." Presented to the World Congress in Sport Science in Golf, in St. Andrews, Scotland.

3. J. N. Vickers (2007) *Perception, Cognition and Decision Training: The Quiet Eye in Action,* Human Kinetics, Europe.

4. S. L. Vine, L. J. Moore, and M. R. Wilson (2011) "Quiet eye training facilitates competitive putting performance in elite golfers," *Frontiers of Psychology,* 2 (8), 1–9.

PAGES 18–19

1. I. J. Deary et al (2009) "Age-associated cognitive decline," *British Medical Bulletin,* 92 (1), 135–152.

2. T. H. Versteegh, A. A. Vandervoort, D. M. Lindsay, and S. K. Lynn (2009) "Fitness, performance and injury prevention strategies for the senior golfer," *International Journal of Sports Science and Coaching,* 3 (1), 199–214.

3. D. M. Lindsay, J. F. Horton, and A. Vandervoort (2000) "A review of injury characteristics, aging factors and prevention programmes for the older golfer," *Sports Medicine,* 30 (20), 89–103.

4. C. J. Thompson, K. M. Cobb, and J. Blackwell (2007) "Functional training improves clubhead speed and functional fitness in older golfers," *Journal of Strength and Conditioning Research,* 21 (1), 131–137.

5. F. I. Craik and N. S. Rose (2012) "Training cognition: Parallels with physical fitness?" *Journal of Applied Research in Memory and Cognition,* 1, 51–52.

6. S. Rovioet et al (2010) "The effect of midlife physical activity on structural brain changes in the elderly," *Neurobiology of Aging,* 31 (11), 1927–1936.

7. P. A. Ades and M. J. Toth (2005) "Accelerated decline of aerobic fitness with healthy aging: What is the good news?" *Circulation.* American Heart Association, 112 (5), 624–646.

8. W. R. Frontera, V. A. Hughes, R. A. Fielding, M. A. Fiatarone, W. J. Evans, and R. Roubenoff (2000) "Aging of skeletal muscle: A 12-year longitudinal study," *Journal of Applied Psychology,* 88 (4), 1321–1326, American Physiological Society.

PAGES 20–21

1. P. Hurrion (2009) "A biomechanical investigation into weight distribution and kinematic parameters during the putting stroke," *International Journal of Sports Science and Coaching* 4, 89–105.

2. T. C. Sells, Y-S. Tsai, J. M. Smoliga, J. B. Myers, and S. M. Lephart (2007) "Strength, flexibility, and balance characteristics of highly proficient golfers," *Journal of Strength and Conditioning Research,* 21 (4), 1166–1171.

PAGES 22–23

1. J. K. Witt and D. R. Proffitt (2005) "See the ball, hit the ball: Apparent ball size is correlated with batting average," *Psychological Science,* 16, 937–938.

2. J. K. Witt, D. R. Proffitt, and W. Epstein (2005) "Tool use affects perceived distance, but only when you intend to use it," *Journal of Experimental Psychology: Human Perception and Performance,* 31, 880–888.

3. J. K. Witt, S. A. Linkenauger, J. Z. Bakdash, and D. R. Proffitt (2008) "Putting to a bigger hole: Performance relates to perceived size," *Psychological Bulletin and Review,* 25 (3), 581–585.

4. J. K. Witt, S. A. Linkenauger, and D. R. Proffitt (2012) "Get me out of this slump! Visual illusions improve sports performance," *Psychological Science,* 23, 397–399.

PAGES 24–25

1. P. Worsfold, N. A. Smith, and R. J. Dyson (2009) "Kinetic assessment of golf shoe outer sole design features," *Journal of Sports Science and Medicine,* 8, 607–615.

2. P. Worsfold, N. A. Smith, and R. J. Dyson (2007) "A comparison of golf shoe design highlights the greater ground reaction forces with shorter irons," *Journal of Sports Science and Medicine,* 6, 484–489.

PAGES 26–27

1. C. Swann, R. Keegan, D. Piggott, L. Crust, and M. F. Smith (2011) "Exploring flow occurrence in elite golf," *Athletic Insight,* 4, 2.

2. K. Evans et al (2012) "Repeatability of three-dimensional thorax and pelvis kinematics in the golf swing measured using a field-based motion capture system," *Sports Biomechanics,* 11 (2), 262–272.

3. D. M. Wolpert and Z. Ghahramani (2000) "Computational principles of movement neuroscience," *Nature Neuroscience,* 3, 1212–1217.

4. T. Okada, K. C. Huxel, and T. W. Nesser (2011) "Relationship between core stability, functional movement, and performance," *Journal of Strength and Conditioning Research,* 5 (1), 252–261.

PAGES 28–29

1. L. Bezzola, S. Mérillat, and L. Jäncke (2012) "The effect of leisure activity golf practice on motor imagery: An fMRI study in middle adulthood," *Frontiers in Human Neuroscience,* 6, 67.

2. J. Milton, A. Solodkin, P. Hlustik, and S. L. Small (2007) "The mind of expert motor performance is cool and focused," *Neuroimage,* 35, 804–813.

3. J. S. Ross, J. Tkach, P. M. Ruggieri, M. Lieber, and E. Lapresto (2003) "The mind's eye: Functional MR imaging evaluation of golf motor imagery," *American Journal of Neuroradiology,* 24, 1036–1044.

4. K. Yarrow, P. Brown, and J. W. Krakauer (2009) "Inside the brain of an elite athlete: The neural processes that support high achievement in sports," *Nature Reviews Neuroscience,* 10, 585–596.

5. J. Baumeister, K. Reinecke, H. Liesen, and M. Weiss (2008) "Cortical activity of skilled performance in a complex sports-related motor task," *European Journal of Applied Physiology,* 104, 625–631.

6. J. E. Birch (2010) "The inner game of sport: Is everything in the brain?" *Sport, Ethics and Philosophy,* 4 (3), 284–305.

PAGES 30–31

1. M. F. Smith and R. Hillman (2012) "A retrospective service audit of a mobile physiotherapy unit on the PGA European golf tour," *Physical Therapy in Sport,* 13 (1), 41–44.

2. G. S. Gluck, J. A. Bendo, and J. M. Spivak (2008) "The lumbar spine and low back pain in golf: A literature review of swing biomechanics and injury prevention," *The Spine Journal,* 8, 778–788.

3. T. M. Hosea and C. J. Gatt (1996) "Back pain in golf," *Clinical Journal of Sports Medicine,* 15, 37–53.

4. M. Adams and W. G. Hutton (1981) "The relevance of torsion to the mechanical derangement of the lumbar spine," *Spine,* 6, 241–248.

5. M. E. Batt (1992) "A survey of golf injuries in amateur golfers," *British Journal of Sports Medicine,* 26, 63–65.

6. A. J. Fradkin, P. A. Cameron, and B. J. Gabbe (2005) "Golf injuries: Common and potentially avoidable," *Journal of Science and Medicine in Sport,* 8 (2), 163–170.

7. H. Sugaya, A. Tsuchiya, H. Moriya et al (1999) "Low back injury in elite and professional golfers: An epidemiologic and radiographic study," in M. R. Farrally and A. J. Cochran (eds), *Science and Golf III: Proceedings of the World Scientific Congress of Golf,* Human Kinetics, Champaign, IL, pp 83–91.

8. D. M. Lindsay, J. F. Horton, and A. A. Vandervoort (2000) "A review of injury characteristics, aging factors and prevention programmes for the older golfer," *Sports Medicine,* 30 (2), 89–103.

PAGES 32–33

1. A. Swain and G. Jones (1996) "Explaining performance variance: The relative contribution of intensity and direction dimensions of competitive state anxiety," *Anxiety, Stress & Coping: An International Journal,* 9 (1), 1–18.

2. S. T. Chamberlain and B. D. Hale (2007) "Competitive state anxiety and self-confidence: Intensity and direction as relative predictors of performance on a golf putting task," *Anxiety, Stress & Coping: An International Journal,* 20 (2), 197–207.

PAGES 34–35

1. R. J. Maughan (2003) "Impact of mild dehydration on wellness and on exercise performance," *European Journal of Clinical Nutrition,* 57 (Suppl 2), S19–S23.

2. G. Szinnai, H. Schachinger, M. J. Arnuad, L. Linder, and U. Keller (2005) "Effects of water deprivation on cognitive-motor performance in healthy men and women," *American Journal of Physiology: Regulatory, Integrative and Comparative Physiology,* 289, R275–R280.

3. M. F. Smith, A. J. Newell, and M. R. Baker (2012) "Effect of acute mild dehydration on cognitive-motor performance in golf," *Journal of Strength and Conditioning Research,* 26 (11), 3075–3080.

4. K. A. Royal, D. Farrow, I. Mujika, S. L. Halson, D. Pyne, and B. Abernethy (2006) "The effects of fatigue on decision making and shooting skill performance in water polo players," *Journal of Sports Sciences,* 24, 807–815.

PAGES 36–37

1. D. L. Neumann and P. R. Thomas (2009) "The relationship between skill level and patterns in cardiac and respiratory activity during golf putting," *International Journal of Psychophysiology,* 72, 276–282.

2. D. L. Neumann and P. R. Thomas (2011) "Cardiac and respiratory activity and golf putting performance under attentional focus instructions," *Psychology of Sport and Exercise,* 12 (4), 451–459.

3. J. J. Bell and J. Hardy (2009) "Effects of attentional focus on skilled performance in golf," *Journal of Applied Sport Psychology,* 21 (2), 163–177.

4. N. Perkins-Ceccato, S. R. Passmore, and T. D. Lee (2003) "Effects of focus of attention depend on golfers' skill," *Journal of Sports Sciences,* 21, 593–600.

CHAPTER 2 the swing

PAGES 40–41

1. J. W. Bunn (1972) *Scientific Principles of Coaching* (2nd ed), Prentice-Hall, Englewood Cliffs, NJ.

2. C. A. Putnam (1993) "Sequential motions of body segments in striking and throwing skills: Descriptions and explanations," *Journal of Biomechanics,* 26 (1), 125–135.

3. P. R. Geisler (2001) "Golf," in E. Shamus and J. Shamus (eds), *Sports Injury Prevention and Rehabilitation,* McGraw-Hill, New York, pp 185–226.

4. P. J. Cheetham, G. A. Rose, R. N. Hinrichs, R. J. Neal, R. E. Mottram, P. D. Hurrion, and P. F. Vint (2008) "Comparison of kinematic sequence parameters between amateur and professional golfers," in D. Crews and R. Lutz (eds) *Science and Golf V: Proceedings of the World Scientific Congress of Golf,* Energy in Motion, Mesa, AZ, pp 30–36.

PAGES 42–43

1. A. Cochran and J. Stobbs (1968) *The Search For The Perfect Swing.* Heinemann, London.

2. N. Betzler, S. Monk, E. Wallace, S. Otto, and G. Shan (2008) "From the double pendulum model to full-body simulation: Evolution of golf swing modelling," *Sports Technology,* 1, 175–188.

3. T. D. Lee, T. Ishikura, S. Kegel, D. Gonzalez, and S. Passmore (2008) "Head-putter coordination patterns in expert and less skilled golfers," *Journal of Motor Behavior,* 4, 267–272.

4. S. Horan and J. Kavanagh (2012) "The control of upper body segment speed and velocity during the golf swing," *Sports Biomechanics,* iFirst, 1–10.

5. R. H. Sanders and P. C. Owens (1992) "Hub movement during the swing of elite and novice golfers," *International Journal of Sports Biomechanics,* 8, 320–330.

6. D. Mann, S. Coombes, M. Mousseau, and C. Janelle (2011) "Quiet eye and the Bereitschaftspotential: Visuomotor mechanisms of expert motor performance," *Cognitive Processing,* 12, 223–234.

PAGES 44–45

1. J. McLean (1992) "Widen the gap," *Golf Magazine,* December, 49–53.

2. M. McTeigue, S. R. Lamb, R. Mottram, and F. Pirozzolo (1994) "Spine and hip motion analysis during the golf swing," in A. J. Cochran and M. R. Farrally (eds), *Science and Golf II: Proceedings of the World Scientific Congress of Golf,* E & FN Spon, London, pp 50–58.

3. P. J. Cheetham, P. E. Martin, R. E. Mottram, and B. F. St. Laurent (2001) "The importance of stretching the 'X-factor' in the downswing of golf: The 'X-factor stretch'," in P. R. Thomas (ed), *Optimising Performance in Golf.* Australian Academic Press, Brisbane, pp 192–199.

4. J. Myers, S. Lephart, Y-S. Tsai, T. Sell, J. Smoliga, and J. Jolly (2008) "The role of upper torso and pelvis rotation in driving performance during the golf swing," *Journal of Sports Sciences,* 26, 181–188.

5. Y. Chu, T. C. Sell, and S. M. Lephart (2010) "The relationship between biomechanical variables and driving performance during the golf swing," *Journal of Sports Sciences,* 28, 1251–1259.

PAGES 46–47

1. A. McHardy and H. Pollard (2005) "Muscle activity during the golf swing," *British Journal of Sports Medicine,* 39, 799–804.

2. S., Marta L. Silva, M.A. Castro, P. Pezarat-Correia, and J. Cabri (2012) "Electromyography variables during the golf swing: A literature review," *Journal of Electromyography and Kinesiology,* 22 (6), 803–813.

PAGES 50–51

1. D. Lindsay, S. Mantrop, and A. Vandervoort (2008) "A review of biomechanical differences between golfers of varied skill levels," in S. Jenkins (ed), *Annual Review of Golf Coaching 2008,* Multi-Science Publishing, Brentwood, pp 187–197.

2. M. McTiegue, S. Lamb, R. Mottram, and F. Pirozzolo (1994) "Spine and hip motion analysis during the golf swing," in A. Cochran and M. Farrally (eds), *Science and Golf II: Proceedings of the World Scientific Congress of Golf,* E & FN Spon, London, pp 50–58.

3. N. Zheng, S. Barrentine, G. Fleisig, and J. Andrews (2008) "Kinematic analysis of swing in pro and amateur golfers," *International Journal of Sports Medicine,* 29, 487–493.

4. S. Barrentine, G. Fleisig, and H. Johnson (1994) "Ground reaction forces and torques of professional and amateur golfers," in A. Cochran and M. Farrally (eds), *Science and Golf II: Proceedings of the World Scientific Congress of Golf,* E & FN Spon, London, pp 33–39.

PAGES 52–53

1. K. Ball and R. Best (2007) "Different centre of pressure patterns within the golf stroke I: Cluster analysis," *Journal of Sports Sciences,* 25 (7), 757–770.

2. S. Jenkins (2008) "Weight transfer, golf swing theory and coaching." in S. Jenkins (ed), *Annual Review of Golf Coaching 2008,* Multi-Science Publishing, Brentwood, pp 29–51.

3. M. Bennett and A. Plummer (2009) *The Stack and Tilt Swing: The Definitive Guide to the Swing That Is Remaking Golf.* Gotham Books, New York.

4. S. Horan and J. Kavanagh (2012) "The control of upper body segment speed and velocity during the golf swing," *Sports Biomechanics,* iFirst, 1–10.

5. J. Myers, S. Lephart, Y. Tsai, T. Sell, J. Smoliga, and J. Jolly (2008) "The role of upper torso and pelvis rotation in driving performance during the golf swing," *Journal of Sports Sciences,* 26 (2), 181–188.

PAGES 54–55

1. M. Kräkel, and D. Sliwka (2004) "Risk taking in asymmetric tournaments," *German Economic Review,* 5 (1), 103–116.

2. M. Zuckerman and D. M. Kuhlman (2000) "Personality and risk-taking: Common bisocial factors," *Journal of Personality,* 68 (6), 999–1029.

PAGES 56–57

1. A. Cochran and J. Stobbs (1968) *The Search For The Perfect Swing.* Heinemann, London.

2. C. Chen, Y. Inoue, and K. Shibara (2007) "Numerical study on the wrist action during the golf downswing," *Sports Engineering,* 10, 23–31.

3. E. Sprigings and S. Mackenzie (2002) "Examining the delayed release in the golf swing using computer simulation," *Sports Engineering,* 5, 23–32.

4. E. Sprigings and R. Neal (2000) "An insight into the importance of wrist torque in driving the golfball: A simulation study," *Journal of Applied Biomechanics,* 16, 356–366.

PAGES 58–59

1. J. P. Broker and M. R. Ramey (2007) "A new method for measuring grip force and its distribution during the golf swing," in S. Jenkins (ed), *Annual Review of Golf Coaching 2007,* Multi-Science Publishing, Brentwood, pp 121–134.

2. F. D. Werner and R. C. Greig (2000) *How Golf Clubs Really Work and How to Optimize their Designs,* Origin Inc, Jackson, WY.

3. D. R. Budney (1979) "Measuring grip pressure during the golf swing," *Research Quarterly,* 50, 272–277.

4. E. R. Komi, J. R. Roberts, and S. J. Rothberg (2008) "Measurement and analysis of grip force during a golf shot," *Proceedings of the Institution of Mechanical Engineers, Part P: Journal of Sports Engineering and Technology,* 222, 23–35.

PAGES 60–61

1. S. Jenkins (2007) "Golf coaching and swing plane theories," in S. Jenkins (ed), *Annual Review of Golf Coaching 2007,* Multi-Science Publishing, Brentwood, pp 1–19.

2. C. L. Vaughan (1981) "A three-dimensional analysis of the forces and torques applied by a golfer during the downswing," in A. Morecki, K. Fidelus, K. Kedzior, and A. Witt (eds), *Biomechanics VII-B,* University Park Press, Baltimore, MD, pp 325–331.

3. R. J. Neal and B. D. Wilson (1985) "3D kinematics and kinetics of the golf swing," *International Journal of Sports Biomechanics,* 1, 60–64.

4. B. Lowe and I. H. Fairweather (1994) "Centrifugal force and the planar golf swing," in A. J. Cochran and M. R. Farrally (eds), *Science and Golf II: Proceedings of the World Congress of Golf,* E & FN Spon, London, pp 59–64.

5. S. Coleman and D. Anderson (2007) "An examination of the planar nature of golf club motion in the swings of experienced players," *Journal of Sports Sciences,* 25, 739–748.

PAGES 62–63

1. R. Mann and F. Griffin (1998) *Swing Like a Pro: The Breakthrough Scientific Method of Perfecting Your Golf Swing,* Broadway Books, New York.

2. M. Turvey (1990) "Coordination," *American Psychologist,* 45 (8), 938–953.

3. M. Latash, M. Levin, J. Scholz, and G. Schöner (2010) "Motor control theories and their applications," *Medicina,* 46 (6), 382–392.

4. B. Abernethy, R. Neal, M. Moran, and A. Parker (1990) "Expert–novice differences in muscle activity during the golf swing," in A. Cochran (ed), *Science and Golf: Proceedings of the First World Scientific Congress of Golf,* E & FN Spon, London, pp 55–60.

5. S. Brown, A. Nevill, S. Monk, S. Otto, W. Selbie, and E. Wallace (2011) "Determination of the swing technique characteristics and performance outcome relationship in golf driving for low handicap female golfers," *Journal of Sports Sciences,* 29 (14), 1483–1491.

6. C. Knight (2004) "Neuromotor issues in the learning and control of golf skill," *Research Quarterly for Exercise and Sport,* 75, 9–15.

7. D. Wolpert and Z. Ghahnramani (2000) "Computational principles of movement neuroscience," *Nature Neuroscience,* 3, 1212–1217.

CHAPTER 3 the equipment

PAGES 68–69

1. These tables are adapted from data collected and analyzed by the makers of the TrackMan™ Launch Monitor, TrackMan a/s, Denmark (*www.trackman.dk*). See *http://wishongolf.com/wp-content/uploads/ 2012/07/TrackMan-Driver-Optimization_2010.pdf*

See also *http://www.golfwrx.com/forums/topic/ 691110-just-got-back-from-my-first-club-fitting/*

PAGES 70–71

1. "Focus: Attack angle," Trackman News, 2 (January 2008), 3–5. Available online at *http://www.trackman.dk/download/ newsletter/newsletter2.pdf*

PAGES 86–87

1. "Insight: Tour professional data," *Trackman News,* 6 (January 2010), 4–6. Available online at *http://www. trackman.dk/download/newsletter/newsletter6.pdf*

2. "Focus: Attack Angle," *Trackman News,* 2 (January 2008), 3–5. Available online at *http://www.trackman.dk/ download/newsletter/newsletter2.pdf*

CHAPTER 4 the environment

PAGES 96–97

1. The simulator software used is PING nFlight (PING, Phoenix, AZ).

PAGES 98–99

1. C. A. Miller and A. G. Davenport (1998) "Guidelines for the calculation of wind speed-ups in complex terrain," *Journal of Wind Engineering and Industrial Aerodynamics,* 74–76, 189–197.

PAGES 100–101

1. R. A. Mehta and J. M. Pallis (2001) "Sports ball aerodynamics: Effects of velocity, spin and surface roughness," in F. H. Froes and S. J. Haake (eds), *Materials and Science in Sports,* CD-ROM edition, TMS, Warrendale, PA, pp 185–197.

2. J. N. Libii (2006) "Design of an experiment to test the effect of dimples on the magnitude of the drag on a golf ball," *World Transactions on Engineering and Technology Education,* 5 (3), 477–480.

3. S. Aoyama (1994) "Changes in golf ball performance over the last 25 years," in M. R. Farrally and A. J. Cochran (eds), *Science and Golf II: Proceedings of the Second World Scientific Congress of Golf,* E & FN Spon, London, p 457.

4. S. Aoyama (2001) "Golf ball aerodynamics: Principles of golf ball aerodynamics – History, aerodynamic basics and common myths," Available online at *http://www. furthereducationlessontrader.co.uk/Mech_Eng_Golf%20 Ball%20Aerodynamics.pdf*

5. S. Aoyama (2010). "A modern method for the measurement of aerodynamic lift and drag on golf balls," in *Science and Golf : Proceedings of the First World Scientific Congress of Golf,* Routledge, New York, p 193.

PAGES 102–103

1. R. I. Marshall and M. R. Lindsay (1990) "A comparative study of the properties of bunker sands from links and inland championship golf courses in Great Britain and Ireland," in A. J. Cochran (ed) *Science and Golf: Proceedings of the First World Scientific Congress of Golf*, E & FN Spon, London, pp 346–351.

2. S. W. Baker, A. R. Cole, and S. L. Thornton (1990) "The effect of sand type on ball impacts, angle of repose and stability of footing in golf bunkers," in A. J. Cochran (ed), *Science and Golf: Proceedings of the First World Scientific Congress of Golf*, E & FN Spon, London, pp 352–357.

PAGES 104–105

1. R. J. Neal and L. M. Hubinger (1990) "The effect of groove shape on the post impact kinematics of golf balls," *Australian Journal of Science and Medicine in Sport*, 22 (2), 39–43.

2. B. B. Lieberman (1990) "The effect of impact conditions on golf ball spin-rate," in A. J. Cochran (ed), *Science and Golf: Proceeding of the First World Scientific Congress of Golf*, E & FN Spon, London, pp. 225–230.

3. D. M. Lindsey, W. Hadi, I. Wright, and A. A. Vandervoort (2008) "Comparison of perceived golf shot performance, upper limb stress and ball flight characteristics between solid and brush fiber hitting mats," in D. Crews and R. Lutz (eds), *Science and Golf V: Proceedings of the Fifth World Scientific Congress of Golf, Energy in Motion*, Mesa, AZ, pp 335–343.

PAGES 106–107

1. F. D. Werner and R. C. Greig (2000), *How Golf Clubs Really Work and How to Optimize Their Design*. Origin Inc, Jackson, WY.

2. A. R. Penner (2002) "The run of a golf ball," *Canadian Journal of Physics*, 80, 931–940.

PAGES 108–109

1. A. S. Weller, C. E. Millard, M. A. Stroud, P. L. Greenhaff, and I. A. Macdonald (1997) "Physiological responses to a cold, wet, and windy environment during prolonged intermittent walking," *American Journal of Physiology: Regulatory, Integrative and Comparative Physiology*, 272 (1), R226–R233.

2. W. D. McArdle, F. I. Katch, and V. L. Katch (2009) *Exercise Physiology: Nutrition, Energy, and Human Performance*. Lippincott Williams and Wilkins, Philadelphia, PA.

3. R. J. Osczevski (1995) "The basis of wind chill," *Arctic*, 48, 372–382.

PAGES 110–111

1. S. J. Haake (1991) "The impact of golf balls on natural turf II: Results and conclusions," *Journal of Sports Turf Research Institute*, 67, 128–134.

2. P. D. Hind, S. W. Baker, T. A. Lodge, J. A. Hunt, and D. J. Binns (1996) "A survey of golf greens in Great Britain I: Soil properties," *Journal of Sports Turf Research Institute*, 71, 9–17.

3. S. W. Baker, P. D. Hind, T. A. Lodge, J. A. Hunt, and D. J. Binns (1996) "A survey of golf greens in Great Britain II: Sward characteristics," *Journal of Sports Turf Research Institute*, 71, 23–30.

PAGES 112–113

1. A. R. Penner (2002) "The physics of putting," *Canadian Journal of Physics*, 80, 1–14.

2. A. M. Streich, R. E. Gaussoin, W. W. Stroup, and R. C. Shearman (2005) "Survey of management and environmental influences on golf ball roll distance," *International Turfgrass Society*, 10, 446–454.

3. P. M. Canaway and S. W. Baker (1992) "Ball roll characteristics of five turfgrasses used for golf and bowling greens," *Journal of Sports Turf Research Institute*, 68, 88–94.

PAGES 114–115

1. The data used for the tables are taken from nFlight (PING, Phoenix, AZ).

CHAPTER 5 coaching with technology

PAGES 118–119

1. Y. I. Abdel-Aziz and H. M. Karara (1971) "Direct linear transformation from comparator coordinates into object space coordinates in close-range photogrammetry," *Proceedings of the Symposium on Close-Range Photogrammetry, Falls Church*, VA, American Society of Photogrammetry, Bethesda, MD, pp 1–18.

2. R. J. Neal and B. D. Wilson (1985) "3D kinematics and kinetics of the golf swing," *International Journal of Sport Biomechanics*, 1, 221–232.

PAGES 122–123

1. H. H. Emmen, L. G. Wesseling, R. J. Bootsma et al (1985) "The effect of video-modeling and video feedback on the learning of the tennis service by novices," *Journal of Sports Sciences*, 3 (2), 127–138.

2. P. C. Van Wieringen, H. H. Emmen, R. J. Bootsma et al (1989) "The effect of video-feedback on the learning of the tennis service by intermediate players," *Journal of Sports Sciences*, 7 (2), 153–162.

3. J. L. Thow, R. Naemi, and R. Sanders (2012) "Comparison of modes of feedback on glide performance in swimming," *Journal of Sports Sciences*, 30 (1), 43–52.

4. D. Backstein, Z. Agnidis, R. Sadhu, and H. MacRae (2005) "Effectiveness of repeated video feedback in the acquisition of a surgical technical skill," *Canadian Journal of Surgery*, 48 (3), 195–200.

5. E. Boyer, R. G. Miltenberger, C. Batche, and V. Fogel (2009) "Video modeling by experts with video feedback to enhance gymnastics skills," *Journal of Applied Behavior Analysis*, 42 (4), 855–860.

6. M. Guadagnoli, W. Holcomb, and M. Davis (2002) "The efficacy of video feedback for learning the golf swing," *Journal of Sports Sciences*, 20 (8), 615–622.

7. C. P. Bertam, R. G. Marteniuk, and M. A. Guadagnoli (2007) "On the use and misuse of video analysis," *Annual Review of Golf Coaching*, 37–46.

8. R. Lorimer and S. Jowett (2010) "Feedback of information in the empathic accuracy of sport coaches," *Journal of Sport and Exercise Psychology*, 11 (1), 12–17.

PAGES 124–125

1. G. R. Watson (1969) *Roman Soldier*. Cornell University Press, New York.

2. J. Knapik (1989) *Load Carried by Soldiers: Historical, Physiological, Biomechanical and Medical Aspects*. Technical Report, US Army Research Institute of Environmental Medicine, Natick, MA.

3. C. Foster, J. Hoyos, C. Earnest et al (2005) "Regulation of energy expenditure during prolonged athletic competition," *Medicine and Science in Sports and Exercise*, 37 (4), 670–675.

4. B. Baron, F. Moullan, F. Deruelle et al (2011) "The role of emotions on pacing strategies and performance in middle and long duration sport events," *British Journal of Sports Medicine*, 45, 511–517.

PAGES 128–129

1. R. Penner (2003) "The Physics of Golf," *Reports on Progress in Physics*, 66, 131–171.

PAGES 132–133

1. A. M. Kleinnijenhuis et al (2008) "Golf performance enhancement and real-life neurofeedback training using personalized event-locked EEG profiles," *Journal of Neurotherapy: Investigations in Neuromodulation, Neurofeedback and Applied Neuroscience*, 11 (4), 11–18.

2. L. Lagos et al (2011) "Virtual reality–assisted heart rate variability biofeedback as a strategy to improve golf performance: A case study," *Biofeedback*, 39 (1).

3. T. E. Sokhadze (2012) "Peak performance training using prefrontal EEG biofeedback," *Biofeedback*, 40 (1), 7–15.

4. F. D. Perry, L. Shaw, and L. Zaichkowsky (2011) "Biofeedback and neurofeedback in sports," *Biofeedback*, 39 (3), 95–100.

5. A. H. A. Razak et al (2012) "Foot plantar pressure measurement system: A review," *Sensors*, 12, 9884–9912.

6. E. A. Murakami et al (2011) "Development of three axis spike force sensor and design of an asymmetric golf shoe outer sole," *Footwear Science*, 3 (1), S117–S120.

7. A. Seaman and J. McPhee (2012) "Comparison of optical and inertial tracking of full golf swings," *Procedia Engineering*, 34.

CHAPTER 6 the practice process

PAGES 136–137

1. J. L. Starkes (1993) "Motor experts: Opening thoughts," in J. L. Starkes and F. Allard (eds), *Cognitive Issues in Motor Expertise,* Amsterdam: Elsevier, pp 3–16.

2. H. A. Simon and W. G. Chase (1973) "Skill in chess: Experiments with chess-playing tasks and computer simulation of skilled performance throw light on some human perceptual and memory processes," *American Scientist,* 61 (4), 394–403.

3. A. K. Ericsson, R. T. Krampe, and C. Tesch-Römer (1993) "The role of deliberate practice in the acquisition of expert performance," *Psychological Review,* 100, 363–406.

4. W. F. Helsen, J. L. Starkes, and N. J. Hodges (1998) "Team sports and the theory of deliberate practice," *Journal of Sport and Exercise Psychology,* 20 (1), 12–34.

5. R. Hayman, R. Polman, J. Taylor et al (2011) "Development of elite adolescent golfers," *Talent Development and Excellence,* 3 (2), 249–261.

6. R. Vaeyens, A. Güllich, C. R. Warr et al (2009) "Talent identification and promotion programmes of Olympic athletes," *Journal of Sports Sciences,* 27 (13), 1367–1380.

7. J. Côté, R. Lidor, and D. Hackfort (2009) "ISSP position stand: To sample or to specialize? Seven postulates about youth sport activities that lead to continued participation and elite performance," *International Journal of Sport and Exercise Psychology,* 7 (1), 7–17.

8. K. A. Ericsson, R. T. Krampe, and S. Heizmann (1993) "Can we create gifted people?" *Ciba Foundation Symposium,* 178, 222–249.

9. B. S. Bloom and L. A. Sosniak (1985) *Developing Talent in Young People,* New York: Ballantine Books.

10. J. Côté and J. Hay (2002) "Children's involvement in sport: A developmental perspective," in J. M. Silva and D. E. Stevens (eds), *Psychological Foundations of Sport,* Boston, MA: Allyn & Bacon, pp 484–502.

11. J. Côté, J. Baker, and B. Abernethy (2003) "From play to practice: A developmental framework for the acquisition of expertise in team sports," in J. Starkes and K. A. Ericsson (eds), *Expert Performance in Sports: Advances in Research on Sport Expertise,* Champaign, IL: Human Kinetics, p 89.

PAGES 138–139

1. M. P. McHugh and C. H. Cosgrave (2010) "To stretch or not to stretch: The role of stretching in injury prevention and performance," *Scandinavian Journal of Medicine and Science in Sports,* 20 (2), 169–181.

2. K. Woods, P. Bishop, and E. Jones (2007) "Warm-up and stretching in the prevention of muscular injury," *Sports Medicine,* 37 (12), 1089–1099.

3. K. A. Moran, T. McGrath, B. M. Marshall et al (2009) "Dynamic stretching and golf swing performance," *International Journal of Sports Medicine,* 30 (2), 113–118.

4. M. H. Anshel and C. A. Wrisberg (1993) "Reducing warm-up decrement in the performance of the tennis serve," *Journal of Sport and Exercise Psychology,* 15, 290–303.

5. A. J. Fradkin, P. A. Cameron, and B. J. Gabbe (2007) "Is there an association between self-reported warm-up behaviour and golf related injury in female golfers?" *Journal of Science and Medicine in Sport,* 10 (1), 66–71.

6. A. J. Fradkin, C. F. Finch, and C. A. Sherman (2001) "Warm up practices of golfers: Are they adequate?" *British Journal of Sports Medicine,* 35 (2), 125–127.

7. R. Ajemian, A. D'Ausilio, H. Moorman et al (2010) "Why professional athletes need a prolonged period of warm-up and other peculiarities of human motor learning," *Journal of Motor Behavior,* 42 (6), 381–388.

8. J. B. Church, M. S. Wiggins, F. M. Moode et al (2001) "Effect of warm-up and flexibility treatments on vertical jump performance," *Journal of Strength Conditioning Research,* 15 (3), 332–336.

9. J. C. Gergley (2010) "Latent effect of passive static stretching on diver clubhead speed, distance, accuracy, and consistent ball contact in young male competitive golfers," *Journal of Strength Conditioning Research,* 24 (12), 3326–3333.

10. A. J. Fradkin, C. A. Sherman, and C. F. Finch (2004) "Improving golf performance with a warm-up conditioning programme," *British Journal of Sports Medicine,* 38 (6), 762–765.

PAGES 140–141

1. P. M. Fitts and M. I. Posner (1967) *Human Performance,* Brooks/Cole, Belmont, CA.

2. A. E. Hernandez, A. Mattarella-Micke, R. W. Redding et al (2011) "Age of acquisition in sport: Starting early matters," *American Journal of Psychology,* 124 (3), 253–260.

3. J. R. Flanagan, P. Vetter, R. S. Johansson et al (2003) "Prediction precedes control in motor learning," *Current Biology,* 13 (2), 146–150.

4. A. Maslow (1963) "Further notes on the psychology of being," *Journal of Humanistic Psychology,* 3, 120–135.

5. L. Bezzola, S. Merillat, C. Gaser et al (2011) "Training-induced neural plasticity in golf novices," *Journal of Neuroscience,* 31 (35), 12444–12448.

6. A. R. Luft and M. M. Buitrago (2005) "Stages of motor skill learning," *Molecular Neurobiology,* 32 (3), 205–216.

7. A. Floyer-Lea and P. M. Matthews (2005) "Distinguishable brain activation networks for short- and long-term motor skill learning," *Journal of Neurophysiology,* 94 (1), 512–518.

8. K. Sakai, O. Hikosaka, and K. Nakamura (2004) "Emergence of rhythm during motor learning," *Trends in Cognitive Sciences,* 8 (12), 547–553.

PAGES 142–143

1. J. Finn (2008) "An introduction to using mental skills to enhance performance in golf: Beyond the bounds of positive and negative thinking," *International Journal of Sports Science and Coaching* 3, 255–269.

2. J. Grezes and J. Decety (2001) "Functional anatomy of execution, mental simulation, observation, and verb generation of actions: A meta-analysis," *Human Brain Mapping* 12 (1), 1–19.

PAGES 144–145

1. J. R. Flanagan, P. Vetter, R. S. Johansson et al (2003) "Prediction precedes control in motor learning," *Current Biology,* 13 (2), 146–150.

2. G. S. Snoddy (1926) "Learning and stability," *Journal of Applied Psychology,* 10, 1–36.

3. E. Crossman (1959) "A theory of the acquisition of speed-skill," *Ergonomics,* 2 (2), 153–166.

4. P. M. Fitts (1964) "Perceptual-motor skill learning," *Categories of Human Learning,* 47, 381–391.

5. A. Newell and P. S. Rosenbloom (1981) "Mechanisms of skill acquisition and the law of practice," in J. R. Anderson (ed), *Cognitive Skills and Their Acquisition,* Erlbaum, Hillsdale, NJ, pp 1–55.

PAGES 146–147

1. S. T. Osis and D. J. Stefanyshyn (2012) "Golf players exhibit changes to grip speed parameters during club release in response to changes in club stiffness," *Human Movement Science,* 31 (1), 91–100.

2. R. A. Schmidt, D. E. Young, S. Swinnen et al (1989) "Summary knowledge of results for skill acquisition: Support for the guidance hypothesis," *Journal of Experimental Psychology: Learning, Memory and Cognition,* 15 (2), 352–359.

3. B. D. Butki and S. J. Hoffman (2003) "Effects of reducing frequency of intrinsic knowledge of results on the learning of a motor skill," *Perceptual and Motor Skills,* 97 (2), 569–580.

4. P. J. Smith, S. J. Taylor S.J., and K. Withers (1997) "Applying bandwidth feedback scheduling to a golf shot" *Research Quarterly for Exercise and Sport,* 68 (3), 215–221.

5. M. Guadagnoli and K. Lindquist (2007) "Challenge point framework and efficient learning of golf," *International Journal of Sports Science and Coaching,* 2, 185–197.

6. J. Baumeister, K. Reinecke, H. Liesen et al (2008) "Cortical activity of skilled performance in a complex sports related motor task," *European Journal of Applied Physiology,* 104 (4), 625–631.

PAGES 150–151

1. J. Karlsen and J. Nilsson (2008) "Distance variability in golf putting among highly skilled players: The role of green reading," *International Journal of Sports Science and Coaching,* 3 (1), 71–80.

2. J. Karlsen, G. Smith, and J. Nilsson (2007) "The stroke has only a minor influence on direction consistency in golf putting among elite players," *Journal of Sports Sciences,* 26 (3) 243–50.

3. D. E. Tierney and R. H. Cooper (1999) "A bivariate probability model for putting proficiency," in M. R. Farrally and A. J. Cochran (eds), *Science and Golf III: Proceedings of the World Scientific Congress of Golf.* Human Kinetics, Champaign, IL, pp 385–394.

4. J. Karlsen (2010) *Performance In Golf Putting,* Oslo: Norwegian School of Sport Sciences.

5. G. J. Kirby (2011) *A Scientist Let Loose on the Golf Green.* Available online at *www.geoffkirby.co.uk/ PuttingScience.pdf*

PAGES 152–153

1. Geoff Mangum's PuttingZone: *Ball-Hole Capture Physics and Optimal Delivery Speed.* Available online at *www. puttingzone.com/capture.html* See also Geoff Mangum's PuttingZone, *Optimal Putting.* Available online at *www. puttingzone.com/optimalputting.html*

2. A. R. Penner (2002) "Physics of putting," *Canadian Journal of Physics,* 80, 1–14.

PAGES 154–155

1. C. P. Bertram, L. Grosser, and M.A. Guadagnoli (2008) "Getting the 'feel' for it: The effects of kinesthetic practice on the golf swing performance," in D. Crews and R. Lutz (eds), *Science and Golf V: Proceedings of the World Scientific Congress of Golf,* Energy in Motion, Mesa, AZ, pp 279–285.

2. S. G. S. Coleman and S. Ritchie (2008) "Mathematical comparison of swing planes with and without the Explanar® Trainer," in D. Crews and R. Lutz (eds) *Science and Golf V: Proceedings of the World Scientific Congress of Golf,* Energy in Motion, Mesa, AZ, pp 263–269.

3. M. W. Kernodle and E. T. Turner (1998) "The use of guidance techniques in the teaching of tennis, badminton and racquetball," *The Journal of Physical Education, Recreation and Dance,* 69 (5) 49–54.

4. N. J. Hodges and P. Campagnaro (2012) "Physical guidance research: Assisting principles and supporting evidence," in N. Hodges and A. M. Williams (eds), *Skill Acquisition in Sport: Research, Theory and Practice,* 2nd ed, Routledge, Abingdon, UK, pp 150–169.

5. P. S. Glazier (2010) "Augmenting golf practice through the manipulation of physical and informational constraints," in I. Renshaw, K. Davids, and G. Savelsbergh (eds), *Motor Learning in Practice: A Constraints-Led Approach,* Routledge, Abingdon, UK, pp 187–198.

6. C. A. Knight (2004) "Neuromotor issues in the learning and control of golf skill," *Research Quarterly for Exercise and Sport,* 75 (1), 9–15.

7. N. Perkins-Ceccato, S. R. Passmore, and T. D. Lee (2003) "Effects of attention depend on golfers' skills," *Journal of Sports Sciences,* 21, 593–600.

8. A. K. Ericsson, R. T. Krampe, and C. Tesch-Römer (1993) "The role of deliberate practice in the acquisition of expert performance," *Psychological Review,* 100, 363–406.

PAGES 156–157

1. W. F. Battig (1956) "Transfer from verbal pretraining to motor performance as a function of motor task complexity," *Journal of Experimental Psychology,* 51 (6), 371.

2. W. F. Battig (1979) "The flexibility of human memory," in L. S. Cermak and F.I.M. Craik (eds), *Levels of Processing in Human Memory.* Hillsdale, N.J., Erlbaum, pp 23–44.

3. J. B. Shea and R. L. Morgan (1979) "Contextual interference effects on the acquisition, retention and transfer of a motor skill," *Journal of Experimental Psychology: Human Learning, Memory and Cognition,* 5 (2), 179–187.

4. L. Bortoli, C. Robazza, V. Durigon et al (1992) "Effects of contextual interference on learning technical sports skills," *Perceptual and Motor Skills,* 75 (2), 555–562.

5. K. G. Hall, D. A. Domingues, and R. Cavazos (1994) "Contextual interference effects with skilled baseball players," *Perceptual and Motor Skills,* 78 (3), 835–841.

6. G. Y. Hwang, D. L. Wright, R. McBride et al (2004) "Experiencing greater contextual interference during practice impacts movement kinematics of the golf putt," *Research Quarterly for Exercise and Sport,* 75 (Supplement A), p 47.

7. G. Y. Hwang (2003) "An examination of the impact of introducing greater contextual interference during practice on learning to golf putt," unpublished Ph.D. thesis, Texas A&M University.

8. J. M. Porter, D. Landin, E. P. Hebert et al (2007) "The effects of three levels of contextual interference on performance outcomes and movement patterns in golf skills," *International Journal of Sports Science and Coaching,* 2 (3), 243–255.

9. J. E. Goodwin and H. J. Meeuwsen (1996) "Investigation of the contextual interference effect in the manipulation of the motor parameter of overall force," *Perceptual and Motor Skills,* 83 (3), 735–743.

10. J. M. Porter and R. A. Magill (2010) "Systematically increasing contextual interference is beneficial for learning sport skills," *Journal of Sports Science,* 28 (12), 1277–1285.

11. M. Guadagnoli and W. Holcomb (1999) "Variable and constant practice: Ideas for successful putting," in A. J. Cochran and M. R. Farrally (eds), *Science and Golf III. Proceedings of the World Scientific Congress of Golf.* E & FN Spon, London, pp 261–270.

12. R. W. Christina and E. Alpenfels (2002) "Why does traditional training fail to optimize playing performance?" in E. Thain (ed), *Science and Golf IV: Proceedings of the World Scientific Congress of Golf.* Taylor Francis, London, pp 231–245.

13. T. K. Dail and R. W. Christina (2004) "Distribution of practice and metacognition in learning and long-term retention of a discrete motor task," *Research Quarterly for Exercise and Sport,* 75 (2), 8.

14. M. G. Wade and H. T. A Whiting (eds) (1986) "Constraints on the development of coordination," in *Motor Development in Children: Aspects of Coordination and Control.* Martinus Nijhoff, Leiden, p 341.

15. T. Schack and D. Hackfort (2007) "Action-theory approach to applied sport psychology," in G. Tenenbaum and R. C. Eklund (eds), *Handbook of Sport Psychology,* 3rd edn. John Wiley & Sons, Hoboken NJ, pp 332–351.

16. J. Hellström (2009) "Competitive elite golf: A review of the relationships between playing results, technique and physique," *Sports Medicine,* 39 (9), 723–741.

17. J. Hellström (2009) "Psychological hallmarks of skilled golfers," *Sports Medicine,* 39 (10), 845–855.

PAGES 158–159

1. M. Sommer and L. Rönnqvist (2009) "Improved motor-timing: Effects of synchronized metronome training on golf shot accuracy," *Journal of Sports Science and Medicine,* 8, 648–656.

2. T. M. Libkuman, H. Otani, and J. Steger (2002) "Training in timing improves accuracy in golf," *Journal of General Psychology,* 129, 17–20.

3. S. L. Bengtsson, F. Ullen, H. H. Ehrsson et al (2009) "Listening to rhythms activates motor and premotor cortices," *Cortex,* 45, 62–71.

CHAPTER 7 the score

PAGES 164–165

1. M. Brodie (2010) "Assessing golfer performance on the PGA Tour," *Interfaces,* 42 (2),146–116. Available online at *www.columbia.edu/~mnb2/broadie/Assets/strokes_ gained_pga_broadie_20110408.pdf*

2. Golf Data Lab, *Why Winners Win* (October 2010). Available online at *www.golfdatalab.com/*

PAGES 166–167

1. A study by Golf Data Lab (*www.golfdatalab.com/*) in which a sample size of almost 1.5 million shots at pro and elite amateur level was examined.

Table of measurements

Distance

1 in = 25.4 mm = 2.54 cm
1 cm = 10 mm = 0.394 in
1 ft = 0.305 m
1 m = 3.281 ft
1 mile = 1.609 km
1 km = 0.621 mile

Speed

1 ft/s = 0.305 m/s = 0.682 mph = 1.097 km/h
1 m/s = 3.281 ft/s = 2.237 mph = 3.6 km/h
1 mph = 1.609 km/h = 1.467 ft/s = 0.447 m/s
1 km/h = 0.621 mph = 0.911 ft/s = 0.278 m/s

Acceleration

1 ft/s^2 = 0.305 m/s^2
1 m/s^2 = 3.281 ft/s^2

Force

1 lb = 4.448 N
1 N = 0.225 lb

Mass*

1 oz = 28.350 g
1 g = 0.0357 oz
1 lb = 453.6 g = 0.454 kg
1 kg = 1000 g = 2.205 lb
* At the surface of the Earth

Converting mass to weight on Earth

1 kg = 2.205 lb = 9.807 N
1 N = 0.225 lb = 0.102 kg

Area

1 cm^2 = 0.155 in^2
1 in^2 = 6.452 cm^2
1 ft^2 = 0.0929 m^2
1 m^2 = 10.764 ft^2

Volume

1 in^3 = 0.0164 liter
1 liter = 61.0237 in^3
1 ft^3 = 0.0283 m^3
1 m^3 = 35.315 ft^3
1 gallon = 3.785 liter
1 liter = 0.264 gallon

Temperature

°F = (°C × 9/5) + 32
°C = (°F − 32) x 5/9
K = °C + 273.15
°C = K − 273.15

Abbreviations

in	inch
ft	foot
mm	millimeter (0.001 m)
cm	centimeter (0.01 m)
m	meter
km	kilometer (1,000 m)
ms	millisecond (1000 s)
s	second
h	hour
y	year
oz	ounce
lb	pound
N	newton
kg	kilogram
g	gram
°F	degrees Fahrenheit
°C	degrees Celsius
K	kelvin
Hz	hertz (cycles per second)

Glossary

absolute temperature The temperature measured in kelvin (K), rather than degrees Fahrenheit (°F) or degrees Celsius (°C). Water freezes at 273.15 K and boils at 373.15 K at standard atmospheric pressure.

accelerometer An instrument for measuring acceleration.

acute angle An angle of less than 90 degrees but more than 0 degrees.

address The act of taking a stance ready to hit the ball, placing the clubhead behind the ball. If the ball moves once a player has addressed the ball, there is a one-stroke penalty, unless it is clear that the act of the player did not cause the ball to move, but a force such as the wind did. Gravity causing it to move upon address is a penalty stroke.

angle of attack (AOA) The angle at which the clubhead strikes the ball. This affects the trajectory the ball will travel and spin.

angular (rotational) velocity A measure of rotational speed; the rate of change of an object's angular position with time.

backspin When a player strikes the golf ball and it flies with a backward spin. The spin gives the ball lift (*see* Magnus force) and makes the ball stop quickly or roll backward on landing.

bounce angle Technically, the measure of the angle from the front edge of a club's sole to the point that rests on the ground when addressing the ball.

carry The distance the ball travels through the air.

characteristic time A measure of the spring-like effect of a golf club—it is the time that a ball remains in contact with the clubface during a stroke, and is directly related to the flexibility of the golf clubhead.

chip A short shot, usually played from very close to the green, which travels a short distance through the air and then rolls the rest of the way to the hole.

clubhead speed A measure, in miles or kilometers per hour, of how fast the clubhead is traveling at the point it impacts the golf ball.

coefficient of friction, μ The ratio of the force required to move two sliding surfaces over each other to the force holding them together. The higher the coefficient, the better the grip.

coefficient of restitution (COR) A measurement of the energy loss or retention when two objects collide. COR is always expressed as a number between 0.000 (meaning all energy is lost in the collision) and 1.000 (which means a perfect, elastic collision in which all energy is transferred from

one object to the other). In golf, clubs collide with golf balls, and the more energy transferred from the club to the ball in that collision, the farther the ball can travel.

control A control in a scientific investigation is an experiment or observation designed to minimize the effects of variables other than the single independent variable. This increases the reliability of the results, often through a comparison between control measurements and the other measurements.

declarative (or explicit) memory One of two types of long-term human memory. It refers to memories that can be consciously recalled, such as facts and knowledge.

degrees of freedom (DOF) The number of independent parameters that determine the state of a physical system, important to the analysis of systems of bodies in mechanical and structural engineering, for example. The position of a rigid body in space is defined by three components of translation and three components of rotation, which means that it has six degrees of freedom.

deliberate practice Consistent directed effort that leads to improved performance, designed with clear objectives and goals.

drag A force acting on a moving body, opposite in direction to the movement of the body, caused by the interaction of the body and the medium it moves through. The strength of drag usually depends on the velocity of the body.

drag coefficient The ratio of the drag on a body moving through air to the product of the velocity and the surface area of the body.

draw A shot that, for a right-handed golfer, curves to the left; often played intentionally by skilled golfers. An overdone draw usually becomes a hook.

dynamic loft The loft (angle) of the part of the club that makes impact with and influences initial direction of the ball, relative to vertical (vertical = zero degrees).

electromagnetism The production of a magnetic field by current flowing in a conductor.

electromyography The recording and study of the intrinsic electrical properties of skeletal muscle. When at rest, normal muscle is electrically silent, but when the muscle is active, an electrical current is generated. In electromyography the electrical impulses are picked up by needle electrodes inserted into the muscle and amplified on an oscilloscope screen in the form of wave-like tracings. The visual recording may be accompanied by auditory monitoring in which the sounds are amplified.

face angle Refers to the position of the clubface relative to the target line. It is measured in degrees and the value is often given on manufacturers' web sites when they list the specifications of their clubs. If the clubface is aligned directly at the target line, the face angle is "square." An "open" face angle means the clubface is aligned to the right of the target line (for right-handed players). If the face angle is "closed," the clubface is aligned to the left of the target line (for right-handers).

fade A shot that, for a right-handed golfer, curves slightly to the right, and is often played intentionally by skilled golfers. An overdone fade will appear similar to a slice.

fast fibers Muscle fibers that produce energy by breaking down glycogen in the absence of oxygen. They produce rapid contractions, but create lactic acid as a by-product.

fat-free mass The combined mass of the body of everything that is not fat (e.g. muscle, bone, skin, and organs).

flop A short shot, played with an open stance and an open clubface, designed to travel very high in the air and land softly on the green. The flop shot is useful when players do not have "much green to work with," but should only be attempted on the best of lies.

functional magnetic resonance imaging (fMRI) An MRI procedure that measures brain activity by detecting associated changes in blood flow.

gear effect So called because when a clubhead twists (rotates) in response to an off-center impact, for the split second that the golf ball and the clubface are in contact, they behave as if they were a pair of meshed gear wheels—when the head moves or rotates in one direction, it will induce the ball to spin in the opposite direction. The closer to the face the clubhead COG is located, the less the gear effect and the less spin is affected, so it is essentially only a feature of driver and fairway wood heads. It can virtually be ignored for irons.

greens in regulation (GIR) A green is considered hit "in regulation" if any part of the ball is touching the putting surface and the number of strokes taken is at least two fewer than par (i.e. by the first stroke on a par 3, the second stroke on a par 4, or the third stroke on a par 5). Greens in regulation percentage is one of many statistics kept by the PGA Tour.

gyroscope A device consisting of a spinning mass, typically a disk or wheel, mounted on a base so that its axis can turn freely in one or more directions and thereby maintain its orientation regardless of any movement of the base.

heel (of club) The part of the head of a golf club where it joins the shaft.

homeostasis The internal balance the body must maintain to ensure health.

horizontal bulge The curvature across the face (side-to-side) of a wood head. It causes the ball to leave the club more to the left or right than a center impact, so results in the ball finishing less off-line as a result of vertical gear effect than if the face were flat.

horizontal gear effect When the ball is struck toward the toe of a driver or fairway wood, the head will twist open and cause the ball to leave the face with draw or hook spin. Heel-side impacts will produce fade or slice spin (*see* gear effect).

hosel The hollow part of the clubhead where the shaft is attached. Hitting the ball off the hosel is known as a shank.

hypokinetic diseases Conditions that arise as a result of a sedentary lifestyle, such as obesity, back problems, high blood pressure, and cardiovascular disease.

inelastic collision A collision in which part of the kinetic energy is changed to some other form of energy.

kinematic chain A group of body segments connected by joints such that the segments operate together to provide a wide range of motion for a limb.

launch angle The initial elevation angle of the ball (with respect to the ground) immediately after impact with the clubhead.

layup A stroke played with a shorter range club than is possible in order to position the ball in a certain spot. This may be done to ensure a more comfortable next stroke or to avoid a hazard.

lie (i) How the ball is resting on the ground, which may add to the difficulty of the next stroke.
(ii) The angle between the center of the shaft and the sole of the clubhead.

lob A lob shot is a relatively vertical (steep arced) shot, usually played with a lofted wedge, intended to land softly and not roll far.

loft The angle between the club's shaft and its face.

long game The aspect of golf involving long shots with woods and low irons.

magnetometer An instrument for measuring the magnitude and direction of a magnetic field.

Magnus force The phenomenon whereby a spinning object flying in a fluid creates a whirlpool of fluid around itself, and experiences a force perpendicular to the line of motion. The curved path of a golf ball (slice or hook) is due largely to the ball's spinning motion (about its vertical axis) and the Magnus force, causing a horizontal force that moves the ball from a straight line in its trajectory. Backspin (upper surface rotating backward from the direction of movement) on a golf ball causes a vertical force that counteracts the force of gravity slightly, and enables the ball to remain airborne a little longer than it would were the ball not spinning: this allows the ball to travel farther than a non-spinning (about its horizontal axis) ball.

maximal aerobic function The maximum capacity of an individual's body to transport and use oxygen during incremental exercise, which reflects the physical fitness of the individual.

mechanoreceptor A sensory receptor that responds to mechanical pressure or distortion.

micro-trauma Small, usually unnoticed injuries caused by repetitive overuse.

molecular weight A measure of the sum of the atomic weights of the atoms in a molecule.

moment of inertia (MOI) A quantity that describes how difficult it is to change the angular motion of an object about a particular axis, i.e. its tendency to resist angular acceleration. For a simple particle, it is the product of its mass and the square of its distance from the axis.

momentum The quantity of motion of a moving body, measured as a product of its mass and velocity.

nerve axon The long threadlike extension of a nerve cell that conducts nerve impulses from the cell body.

neurophysiology The branch of physiology that deals with the functions of the nervous system.

normal force The force on an object perpendicular to the surface it rests on, i.e. the force that keeps two surfaces touching, but does not let them come apart or press further together.

parasympathetic nervous system Part of the nervous system whose main function is to conserve/restore your body's energy. For example, it is responsible for sending signals to slow your heart rate and breathing, and speed up your digestive tract in order to digest calories and save energy (*see* sympathetic nervous system).

path angle The angle at which the club hits the ball, which determines the direction of the golf ball at impact.

pitch A short, accurate shot, usually played with a less-than-full swing, using a club with a large loft angle.

pitch mark A divot on the green caused when a ball lands. Players must repair their pitch marks, usually with a tee or a divot tool.

plug A plugged lie is one where the ball is at least half-buried, also known as a "buried lie," or a "fried egg" when it occurs in a bunker.

psychophysiology The branch of physiology dealing with the relationship between physiological processes and thoughts, emotions, and behavior.

putts per round (PpR) Defined by the PGA Tour as the average number of putts per round played.

"quiet eye" moments Used to describe how golfers keep their gaze on the ball absolutely still just before and as the stroke is performed.

sand save When a player makes par by holing the ball in two shots from a green-side bunker.

scramble When a player does not find the green, but still achieves par or better on a hole.

shear force A force acting in a direction parallel to a surface or to a planar cross-section of a body.

short game Play that takes place on or near the green, such as putting, chipping, pitching, and green-side bunker shots.

slice A shot that initially takes a trajectory on the same side of the golf ball from which the player swings but eventually curves sharply back, opposite to the player.

sole (of club) The bottom surface of the head of a golf club.

spin loft The difference between dynamic loft and attack angle, which determines the amount of spin on the ball.

sweetspot The location on the clubface where the optimal ball-striking results are achieved. The closer the ball is struck to the sweetspot, the higher the power transfer ratio will be. Hitting it in the sweetspot is also referred to as hitting it in the screws.

sympathetic nervous system
The sympathetic nervous system tends to act in opposition to the parasympathetic nervous system, for example by speeding up the heartbeat and causing contraction of the blood vessels. It regulates the function of the sweat glands and stimulates the secretion of glucose in the liver. It activated especially under conditions of stress (see parasympathetic nervous system).

toe (of club) The part of the head of a golf club farthest from the shaft.

torque A "twisting force," which causes a change in the rotational motion of a body about an axis.

trajectory The path of a projectile or other moving body through space.

up and down Starting from off the green, the player holes the ball in two strokes—the first gets the ball "up" onto the green, and the second (a putt) gets the ball "down" into the hole.

vector A quantity, such as velocity, completely specified by a magnitude and a direction.

vertical gear effect (VGE) When the ball is struck above the center of gravity (COG) of a driver head, the head will rotate backwards about its COG. If the face had zero loft, that would produce forward spin (topspin), but because ALL drivers have some loft, the effect can only ever be to reduce the amount of backspin. When a ball is struck below the COG the reverse happens and more backspin is produced. Drivers heads will exhibit more VIE than fairway woods, because their deeper faces allow the ball to be struck more above or below the COG than fairway woods (see gear effect).

vertical roll The vertical (top-to-bottom) curvature of the face of a wood head. It causes the ball to leave the clubface higher or lower than a center impact, so can offset the higher or lower spin produced by vertical gear effect.

vestibular system Detects motion of the head in space and in turn generates reflexes that are crucial for our daily activities, such as stabilizing the visual axis (gaze) and maintaining head and body posture. The vestibular system also provides us with our subjective sense of movement and orientation in space.

X-factor The differential in degrees of rotation between the pelvis and thorax during the golf swing.

Notes on contributors

EDITOR

Dr. Mark F. Smith is principal lecturer at the University of Lincoln in the UK. He is a sports scientist with more than 15 years' experience in exercise physiology research and applied consultancy practice. Having worked with a number of professional players, he has made significant contributions to the emerging importance of physiology in the development of golf performance. He has published golf science research in a range of leading international scientific publications, has worked as a scientific advisor for leading organizations, and is on the International Editorial Board for the World Scientific Congress of Golf Science.

CONTRIBUTORS

Andrew Collinson

Beginning his career in golf in 1998 as an Assistant PGA Coach, Andrew developed a passion for golf science, specializing in biomechanics, during the completion of his Sports Science Masters Degree. Applying scientific principles to help coach players, he now works as a biomechanical analyst working with players ranging from beginners, to elite amateurs, to touring professionals. Andrew embarked on a doctorate in 2009 to explore the biomechanics of the golf swing at the University of Lincoln.

Dr. Paul Glazier

Working as a consultant sport and human movement scientist with over 15 years of academic, technical, and practical experience, Dr. Glazier has provided sport science support services to a wide range of athletes and sports teams, from regional juniors to Olympic and World Champions, in a variety of sports, including golf. An expert in sports biomechanics, motor control, skill acquisition, and sports performance analysis, he has published over 40 peer-reviewed journal articles, book chapters, and conference papers in these areas and is currently a Research Fellow at the Institute of Sport, Exercise and Active Living at Victoria University in Melbourne. He has won several national long driving championships, has competed at the RE/MAX World Long Drive Championship Finals, and currently plays off a handicap of 4.

Dr. John Hellström

John has over 20 years of experience as a PGA coach, golf science consultant, and applied researcher. He has recieved the PGA Teacher of the Year Award and has a PhD in Sports Science. As Director of Research and Development at the Swedish Golf Federation, John has published his golf science research within internationally recognized publications. He is the creator of www.GolfPyramid.com, which probably is the most advanced online tool for golfers and coaches. John is also an advisory board member for the Annual Review of Golf Coaching and the Journal of Applied Golf Research, as well as being consultant to several national teams, elite players, and coaches.

Richard Kempton

Richard has more than 16 years experience fitting, building, customizing, and repairing golf clubs. Winner of several awards for his skills and contributions to clubfitting and clubmaking (including the prestigious "European Clubmaker of the Year" title in 2001 from the US-based Golf Clubmaker's Association), he has worked and continues to work with golfers of all abilities from beginners to past winners of the Brabazon Trophy & English Amateur Championship and members of the Men's European, Senior & Satellite Tours and Ladies' European Tour.

Dr. Peter Lamb

Peter Lamb is associate researcher at the Technische Universität München in Munich, Germany and head of the faculty's Golf Biomechanics Laboratory. His research aims at understanding the nature of the complex coordinated action of the golf swing. With expertise in sports biomechanics, motor control, and the use of novel analysis techniques, Peter has acted as scientific consultant for professional golf coaches from Canada, Germany, New Zealand, and the UK. In addition to being a life-long golfer, Peter is also an enthusiastic cyclist, skier, and ice-hockey player.

Graeme Leslie

Graeme Leslie is the Managing Director of Golf Performance Analysis Ltd. and founder of the Performance Measurement system known as Golf Data Lab. This is used by the Scottish and English Golf Unions, the Irish Ladies Golf Union, and a number of overseas associations including US colleges. Professional users play on the European Tour, the US PGA Tour, the Asian Tour, the European Challenge Tour, as well as a number of satellite and regional tours. His dedication to improving understanding of the game through proper performance measurement has led to a consultancy role with the R&A since May 2010. Married with four children and living in the north-east of Scotland, Graeme remains a golf enthusiast, as he has been for over 40 years.

Dr. Robert Neal

Robert is considered a world leader in the application of biomechanics to golf performance. With a prolific research track record lasting over 30 years, Robert currently is Biomechanics Board Member for the Titleist Performance Institute and author of the *Biomechanics Manual for Golf Professional Trainees*. Having published over 20 international scientific articles on golf mechanics, Robert was the first to publish work on 3D kinematic and kinetic analysis of the golf swing. As co-director and co-founder of Golf BioDynamics, a company formed to provide expert 3D golf swing analysis to the golf instructor/golfer, Robert has acted as a consultant for the Danish, English, and German national teams, and supports international touring players on the LPGA, PGA, Australasian, and European Tours.

Dr. Sandy Willmott

Dr. Willmott is a Senior Lecturer at the University of Lincoln who specializes in applying mechanical principles to swinging motions in sport. His research initially focused on field hockey technique but has more recently expanded to include golf. He is an expert reviewer for international publications such as the *Journal of Sports Sciences, Sports Biomechanics*, and the *International Journal of Sports Science and Coaching*.

Index

Acknowledgments

The success of this book is a result of a dedicated team of talented individuals, all of whom have come together to create what I consider to be the most compelling, authoritative book on the science of golf ever written. From around the globe, leaders in their own rights have drawn upon their expertise and the vast scientific research to produce this thoroughly comprehensive scientific compendium of golf.

First and foremost I owe an immense debt to all of the contributors; Paul, Peter, Richard, Sandy, Andrew, Rob, John, and Graeme, for the tireless efforts bringing the book together. All, in their own way, have provided support, talked things over, read, wrote, designed, offered comments, and assisted in the editing, proofreading, and design. I would also like to thank Tom Wishon (Tom Wishon Golf Technology), Dr. Paul Wood (Ping, US) and Chris Sells (StrokeAverage.com) for providing early ideas and feedback as the book took shape.

I would like to express my deepest gratitude to the entire team at Ivy Press, who from the very start allowed me to evolve the concept and take ownership of the project. To Jason Hook, Tom Kitch, and especially Caroline Earle—thank you for your support, guidance, and for believing in me. An indebted thank you must be made to Rob Yarham, our technical editor, who without his expert ability to rework those spreads into something quite extraordinary this project would have ended up in the lake on the first! I would also like to thank Rolo for his illustrative wizardry in creating some amazing graphics throughout the book.

Above all I want to thank my wife, Claire, my children and the rest of my family, who supported and encouraged me in spite of all the time it took me away from them.

Mark F. Smith

The Ivy Press would like to thank the following for permission to reproduce copyright material:
Getty Images: 5, 9, 33, 55, 85, 91, 109; US PGA Tour: 169.
Karen A. Harrison, Golf BioDynamics Inc. (www.golfbiodynamics.com): 117 (photograph of Lara Katzy).
Shutterstock/Vladimir Badaev: 161; Cindy Hughes: 135; Photogolfer: 39; Mai Techaphan: 13; Zorandim: 65.

Every effort has been made to trace copyright holders and to obtain their permission for the use of copyright material.
The publisher apologizes for any errors or omissions in the lists above and will gratefully incorporate any corrections in future reprints if notified.

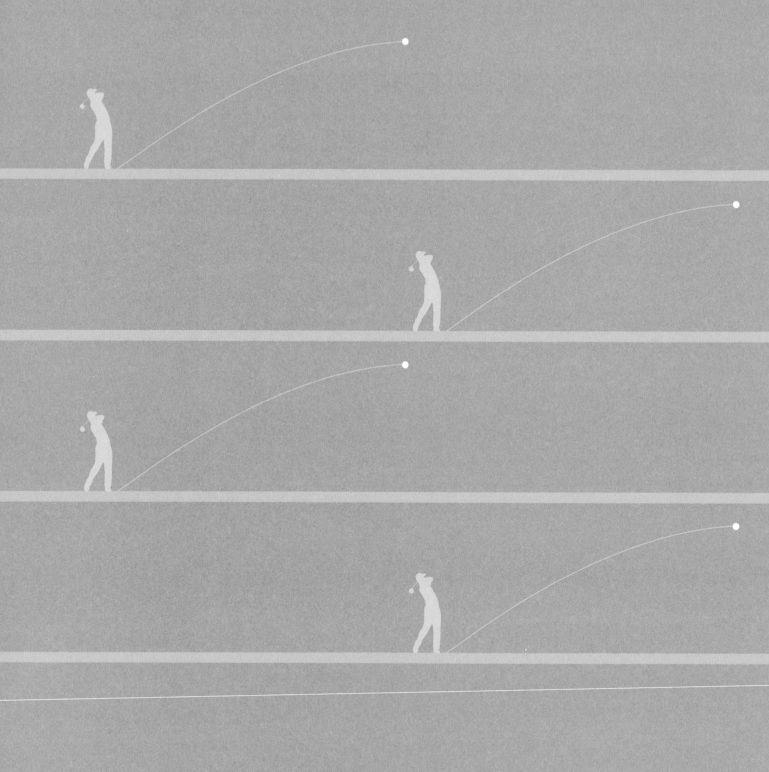